T0135483

Formal Synthesis of Safety Controller Code for Distributed Controllers

Dissertation

zur Erlangung des akademischen Grades

Doktoringenieur (Dr.-Ing.)

vorgelegt dem

Zentrum für Ingenieurwissenschaften der

Martin-Luther-Universität Halle-Wittenberg

als organisatorische Grundeinheit für Forschung und Lehre im Range einer Fakultät

(§75 Abs. 1 HSG LSA, §19 Abs. 1 Grundordnung)

von

Herrn Dipl.-Ing. Dirk Missal

geboren am 16. Dezember 1977 in Halle (Saale)

Gutachter

1. Prof. Dr.-Ing. Hans-Michael Hanisch

2. Prof. Dr. Zhiwu Li

Halle (Saale), den 24. 06. 2011

Reihe: Hallenser Schriften zur Automatisierungstechnik
herausgegeben von:
Prof. Dr. Hans-Michael Hanisch
Lehrstuhl für Automatisierungstechnik
Martin-Luther-Universität Halle-Wittenberg
Theodor-Lieser-Str. 5
06120 Halle/Saale

email: Hans-Michael.Hanisch@informatik.uni-halle.de

Bibliographic information published by the Deutsche Nationalbibliothek

The Deutsche Nationalbibliothek lists this publication in the Deutsche
Nationalbibliografie; detailed bibliographic data are available in the
Internet at http://dnb.d-nb.de

(Hallenser Schriften zur Automatisierungstechnik; 9)

Logos Verlag Berlin GmbH
Comeniushof, Gubener Str. 47,
10243 Berlin
Tel.: +49 030 42 85 10 90
Fax: +49 030 42 85 10 92
INTERNET: http://www.logos-verlag.de

Preface

This work developed mostly during my work for the Chair for Automation Technology at the Institute of Computer Science of the Martin Luther University Halle-Wittenberg. I worked there on distributed controller design, formal modeling and formal synthesis of controllers. The projects were founded by the Deutsche Forschungsgemeinschaft DFG and supported by a number of persons. I want to thank those at the beginning of this work.

My first thank I owe to my supervisor Prof. Dr. Hans-Michael Hanisch. His trust and willingness to facilitate my work on the area of formal controller synthesis was essential for this work. This work would not have been possible unless his specialist guidance support as well as his motivation to push this work to success.

I thank Prof. Dr. Zhiwu Li for his interest on my work and his willingness to provide his second advisory opinion.

Many thanks go to my former colleagues at the chair of automation technology for the inspiring atmosphere and many interesting discussions. Special thanks to Dr.-Ing. Martin Hirsch for interesting talks opening the view for related topics and Dipl.-Ing. Sebastian Preusse for his language related hints and corrections. Further the collaboration with Prof. Dr. Victor Dubinin gave a lot new impulses especially for the formal descriptions in this work.

This thesis would not have been possible unless the support and the consideration of my wife Carina and my parents and the balancing demand of my sons.

June 2011   Dirk Missal

Contents

List of Figures

List of Tables

1 Introduction

Modern control systems in manufacturing are characterized by rising complexity in size and functionality. They are highly decentralized and constitute a network of physically and functionally distributed controllers collaborating to perform the control tasks. That goes along with a further growing demand on safety and reliability.

The development to more complex systems leads to the need of systematic, more reliable design methods on one hand and intensifies the issue of computational complexity for such methods on the other hand. Within the wide range of approaches for systematic controller design, this work focuses on their model-based design. More detailed, approaches for formal synthesis of control functions and their implementation are discussed in the following.

Theoretical approaches for the synthesis of controllers or supervisors are adopted to answer the newly emerging challenges of distribution and communication. This includes topics like the use of modular plant models that preserve information about the modular structure of the plant, methods to specify forbidden or desired behavior locally instead of using global specifications and synthesis approaches for local controllers and their communication structure instead of a global controller or supervisor.

1.1 Model-driven controller design

Formal controller design methods are based on formal behavior models. The formal character of such models allows the use of mathematical operations and rules to analyze or modify them. The result of formal model-based methods are a formal controller model and information about its properties. Formal statements or models are definite while non-formal or semi-formal statements have to be interpreted. Hence, the correctness of statements and operations on formal models can be analyzed and proven. Further, all model-based methods have the advantage that the system behavior can be analyzed without influencing the real system, without complete observation of a real system and with useful abstraction from reality.

On the other hand, model-based design methods add specific issues as there are - amongst others - the cost for the modeling itself, the challenge of finding a suitable application abstraction, the mathematical complexity of the analysis and the assignability of the results back to real systems. All these points are discussed for the introduced approaches in this work.

The term formal model subsumes a wide range of model types and different application areas. There are three main groups of models. First are discrete models describing behavior-based on system states and state transitions. The second group are continuous models representing behavior in terms of differential equations as for example linear system equations. A combination of those two are hybrid models. They combine system's state representation with continuous behavior models enabled in an active state. A simplified overview on model types is given in Figure 1.1. The model used in this work is based on Petri net elements. The higher Petri nets are highlighted therefore.

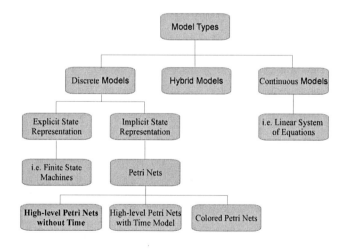

Figure 1.1: Formal model types overview.

Formal model-based analysis and design methods subsume the simulation with formal models, formal verification and formal controller synthesis. The formal model is used to test or respectively verify the behavior of a controller model in closed loop with a model of the plant behavior   in   simulation   and   verification.

The used controller is designed manually in most cases. The idea of formal controller synthesis is to generate a controller model automatically from a model of the uncontrolled plant behavior and a given specification of desired or forbidden behavior. Any formal synthesis approach proven to be correct leads automatically to a correct controller model according to the given specification. The executable controller code should be generated automatically from the controller model in the next step. The steps of controller synthesis are shown in Figure 1.2. The controller synthesis for distributed control functions and their implementation are addressed in this work. The approaches discussed in the following are based on a discrete model with implicit state representation and are exemplified on a manufacturing example.

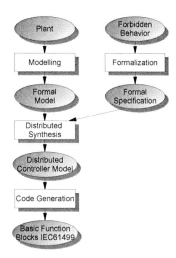

Figure 1.2: Steps to be proceeded for controller function synthesis.

1.2 Problem and methods of resolution

The area of controller design in general and controller synthesis in particular has shown a strong movement during the last years. New potential for application of distributed controllers is driven by software and hardware development.

First fundamental work on controller synthesis was published by the group of P.J. Ramagde and M.W. Wonham introducing the supervisory control theory [WR84, RW87b]. Supervisors are controllers coordinating different machines or machine parts to avoid undesirable behavior or/and deadlocks. Last is also addressed in works on deadlock prevention in so called flexible manufacturing systems [BK90, LZ09, ZL10].

Controlling the computational complexity of synthesis approaches has been the main issue from the beginning [Sre96]. Modular enhancements were developed shortly after introduction of the supervisory control theory [RW87a, LW88] to address the issue of computational complexity. An overview of Discrete Event Systems (abbr. DES) and control theory is given in [RW89]. For all those work it is common that it is based on finite state machine models or language representations [CDFV88]. First higher Petri net model-based synthesis approaches are developed at nearly the same time [Kro87, HK90]. A combination of avoiding forbidden states and inducing a desired process is described in [VH03]. The problem of forbidden state control using non-safe Petri nets is discussed in [DA10]. The authors use a similar specification as applied in this work and derive also boolean descriptions as control functions. The approach is described for a monolithic control problem and requires the computation of the reachable state space.

There are major differences between modular and distributed or decentralized designs. Modular approaches try to divide the control problem into sub-problems to reduce the computational complexity. The modularity is driven by the structure of the control problem. In contrast, the distribution in distributed or decentralized controller synthesis is driven by the aimed control structure. That control structure can consist of more or less interactive control components and is related to the physical distribution of the controlled system. Different possible structures are discussed in Section 3.1.

The specification of supervisors is mainly provided in terms of forbidden states or sequences given in terms of a language. That is common with this work using forbidden state specifications. But the safety control functions in this work are designed on the shop-floor level while the main work in control theory addresses the more abstract supervisor control level. Some preliminary work was published by the group of H.-M. Hanisch and cooperative other groups [HLR96, DHJ+04, DHJ+04, JN04] treating the controller synthesis for that control level. This work is based on the preliminary work of that group and discusses the major issues of achieving distributed controller model results, reducing the computational complexity and proving important properties of the results. Some of the principles introduced in the supervisory control theory framework are adopted for that purpose. These principles are discussed more detailed in the related sections. Two different approaches for the synthesis of distributed safety control functions are discussed in the following chapters. The work therefore is structured as follows.

The used discrete event model type is described in the next Chapter 2 on Safe Net Condition/Event Systems (abbr. $_sNCES$). A new modular semantic for $_sNCES$ is introduced to be able to take advantage of the modular model structure in the analysis. Further, new structural properties are defined extending the theory of $_sNCES$.

Chapter 3 addresses the background of safety control functions and explains how they fit into the general control structures. The distributed control structure is introduced and compared to other distribution approaches discussed in the literature.

The subsequent chapters discuss the synthesis in the sequence given in Figure 1.2. The system example is introduced in Chapter 4. It is a manufacturing system in lab-scale with mainly binary actuators and sensors. The example is used to describe the modeling based on the introduced modeling guidelines. The modeling owns attention because every synthesis result can only be as good as the model is. The behavior specification and its distribution to modular specifications is discussed in Chapter 5. The specification is given in terms of state predicates subsuming a number of markings. It is described how the use of state predicates reduces the complexity of the analysis.

The next two Chapters introduce two approaches for distributed safety control function synthesis. A modular approach is introduced in Chapter 6. It combines the idea of modular analysis with the aim to generate distributed controller models. The modular execution reduces the computational complexity and leads directly to distributed results. The homogeneous approach introduced in Chapter 7 focuses on improved permissiveness of the results and disclaims on modular execution to reach that. This chapter describes the distribution of the synthesized homogeneous controller model to distributed control functions.

The implementation of synthesized controller models in terms of executable controller code is rarely addressed in literature about the supervisory control theory. This work closes that gap for the distributed safety control functions and introduces a method to implement the model in terms of Function Blocks according to the international standard IEC 61499 in Chapter 8. The fit of those generated Function Blocks into the general control implementation is also discussed in this chapter.

This work closes with conclusions of the presented topics in Chapter 9. Further, this chapter gives an outlook on thinkable applications and further work to be done to get the theory into practice.

2 Safe Net Condition/Event Systems

Every model-driven controller design methodology is affected significantly by the chosen model with its particular capabilities. The synthesis approaches for distributed controller components described in the following chapters are specified for a subclass of the Net Condition/Event System (abbr.: $NCES$) models. The NCES models support the idea of synthesis in two major criteria. First, it gives a modular and hierarchical structure. The cost of modeling a system can be reduced by the use of modular and hierarchical composition of system models from basic model parts. The model design is described in detail in Section 4. Second, the modular model structure provides structural information for the distribution of controllers (Section 7) and allows the use of modular analysis algorithms (Section 6).

In this chapter, the syntax and semantic of *safe Net Condition/Event Systems* (abbr.: $_sNCES$) are defined. A more detailed description of the models subsumed under the term $NCES$ and the structure of the dialects is given in [Kar09].

The $NCES$ model was introduced by Hanisch and Rausch [HR95], and it combines behavior encapsulation in modules with a concept of signals for interaction between those modules. The signal concept defines in sections constant condition signals and pointwise constant event signals. Hanisch and Roch combined that signal concept with a behavior model based on Petri net elements. A combination of that signal concept with automata behavior models is described by Krogh and Kowalewski in [KK96]. The modular concept of Hanisch and Rausch includes also hierarchical structures in terms of composite modules. The resulting model syntax is defined in Section 2.1. The model semantic was defined in a not strictly modular way. A modular model semantic is required to work out modular analyzing algorithms for $NCES$ models. Therefore, a refined modular semantic for *safe NCES* is introduced in Section 2.2. The model behavior can be analyzed module by module or within partially composed models based on that modular semantic definition. Composition stands for the formal transformation rules in this context. The rules for composition are defined in Section 2.3. Their application modifies the hierarchical and modular structure through module concentration. In Chapter 6 a modular synthesis

algorithm is introduced. It takes advantage of the modular steps on partially composed
models. Complete composition of models is used for the synthesis in terms of a monolithic
approach in Chapter 7.

Some structural analysis methods on $_sNCES$ are discussed in the last part of this chapter.
Structural analysis has the advantage to mostly be of lower computational complexity than
behavior analysis methods. The introduced invariants are used to exclude some behavioral
elements from analysis. Thus, the fundamentals for the subsequent chapters are given by
the model description in the following sections.

2.1 Syntax of Safe Net Condition/Event Systems

The safe Net Condition/Event System model is a special case of $NCES$. Thereby, no
capacity and arc weights are explicitly defined. The model is 1-bounded by definition and
therefore called *safe*. That means every arc carries one token and every place can hold one
token at most. The limitation to a marking of maximal one for every place is chosen to
reduce the complexity of the analysis algorithms. The approaches have to be generalized
if it is necessary to represent a considered plant by a model with a more general model of
the $NCES$ class. Despite that constraint, a wide range of systems can be modeled using
the modularity and the signal interconnections.

The following definitions of the structure of $_sNCES$ are deduced from the definitions of the
more general and timed $TNCES$ model in [Kar09, Thi02]. Time as well as capacities, arc
weights and all related definitions are not used in the context of this work. The notations
in the following definitions comply with those in [Kar09].

As mentioned, the behavioral elements of the model $_sNCES$ consist of the basic elements
of Petri nets extended by two kinds of signals. Therefore, those elements are defined first.

Definition 2.1.1. (Petri- Net)
A net PN is given by a tuple $PN = (P, T, F, m_0)$ with
P is a finite, non empty set of *places*,
T is a finite, non empty set of *transition* (with $P \cap T = \emptyset$),
$F \subseteq ((P \times T) \cup (T \times P))$ is the *flow relation* (also called a set of *arcs*) and
$m_0 : P \rightarrow \{0, 1\}$ is an initial marking of PN.

$\square$

Petri nets have a well-defined structure und operational semantic. Hence, they form an
achievable behavior model and are widely used in modeling of multifaceted types of systems.

The global system state is represented implicitly by local markings in contrast to the global states specified in automata theory.

Those well-known elements are extended by two kinds of signals. They influence the system dynamic without carrying any tokens. The condition signals establish relations between local markings of a Petri net or different depending Petri nets in a model. The in points constant event signal carries state transition information and constitutes a unilateral synchronization of transitions firing. The nature of the signals within the net structures is the same as for the later introduced signal couplings between modules. Their effect is implicitly defined by the firing rules defined in Section 2.2. The resulting net structure is called $_sNCE - Structure$ and defined as follows:

Definition 2.1.2. ($_s$NCE-Structure)

A safe Net- Condition/Event structure is a tuple

$$\mathcal{S} = \{P, T, F, CN, EN, em, sm\}$$

where

P	is a finite, non empty set of *places*,
T	is a finite, non empty set of *transition* (with $P \cap T = \emptyset$),
$F \subseteq ((P \times T) \cup (T \times P))$	is the *flow relation* (also called a set of *arcs*),
$CN \subseteq (P \times T)$	is the set of *condition arcs*,
$EN \subseteq (T \times T)$	is the set of *event arcs*,

EN is cycle free, i.e.:

1. $\nexists(t_1, t_2) \in EN : t_1 = t_2$ and

2. $\nexists((t_1, t_2), \cdots, (t_{i-1}, t_i)) : (t_{l-1}, t_l) \in EN$ for $2 \leq l \leq i \wedge (t_1 = t_i)$,

$em : T \rightarrow \{\boxed{\wedge}, \boxed{\vee}\}$ assigns an *event mode* to every transition,

$sm : T \rightarrow \{\texttt{i}, \texttt{s}\}$ assigns a *firing mode* to every transition.

□

The event mode for a transition declares whether incoming event signals at a transition are combined disjunctively ("OR") or conjunctively ("AND"). The conjunctive mode is default and the corresponding symbols are omitted. The effect on the model dynamic is described in Section 2.2. The same holds for the firing mode of a transition. The default firing mode is spontaneous "s" and the symbol is omitted for this mode.

The sets P, T and F in a $\mathcal{S}$ are interpreted and represented graphically in the same way as done in Petri nets. Nevertheless, in general they do not form a Petri net when joined with an initial marking m_0. That is because the tuple $\{P, T, F\}$ in $\mathcal{S}$ does not necessarily form a net. Single places without any connected arc as well as single transitions have an interpretation in $_sNCES$ in contrast to Petri nets. Places can be source of condition signals while transitions can be source and/or sink of event signals. Hence, both affect the model behavior by signals. In Petri nets such single nodes are omitted, because they do not have any behavioral impact. If properties or rules are to be transcribed from Petri nets to $_sNCES$, then it has to be verified, if they hold for single nodes too. The tuple $\{P, T, F, m_0\}$ of an $_sNCES$ is called *underlying Petri net* in the following.

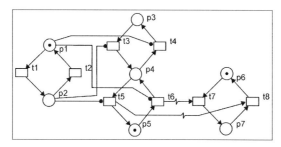

Figure 2.1: Example for an $_sNCE$ structure.

Figure 2.1 shows an example of an $_sNCE$ structure. The event signals are symbolized by an arc with flash and arrow. The arc from t6 to t7 is an event arc example. Condition signals differ from ordinary arcs only by the filled circle at the end. The arc from p1 to t4 is a condition arc.

Notations are declared for the regions describing the predecessor and successor relations of the two kinds of nodes within a structure (based on [Thi02]) as:

$^F t = \{p \in P | (p, t) \in F\}$ and $^{CN} t = \{p \in P | (p, t) \in CN\}$ denotes the *pre-region of a transition t*.

$t^F = \{p \in P | (t, p) \in F\}$ denotes the *post-region of a transition t*, respectively.

Subsets of places are specified by those regions. The regions formed by subsets of the set of transitions are:

$^F p = \{t \in T | (t, p) \in F\}$ denotes the *pre-region of a place p*.

$p^F = \{t \in T | (p, t) \in F\}$ denotes the *post-region of a place p*.

The relations between transitions can be classified in terms of the layout of event signals.

Definition 2.1.3. (Event Sources, Event Sinks, Trigger Transitions, [Thi02])

Let $\mathcal{S}$ be an $_sNCE - structure$ and $t \in T$ a transition.

t is called *event source* iff $\exists t' \in T : (t, t') \in EN$,

t is called *event sink* iff $\exists t' \in T : (t', t) \in EN$.

$^{EN}t = \{t' \in T | (t', t) \in EN\}$ is the set of *event sources* of t, also called the *event pre-region* of the transition.

$t^{EN} = \{t' \in T | (t, t') \in EN\}$ is the set of *event sinks* of t, also called the *event post-region* of the transition, respectively.

t is a *trigger transition* iff $^{EN}t = \emptyset$.

$T_{Ei} := \{t \in T | ^{EN}t = \emptyset \wedge sm(t) = i\}$ is *the set of instantaneous trigger transitions* in $\mathcal{S}$.

$T_{Es} := \{t \in T | ^{EN}t = \emptyset \wedge sm(t) = s\}$ is *the set of spontaneous trigger transitions* in $\mathcal{S}$.

$T_S := \{t \in T | ^{EN}t \neq \emptyset\}$ is *the set of event sinks* in $\mathcal{S}$.

$\square$

Every transition without an incoming event signal is called *trigger transition*. Transitions with at least one incoming event signal are called *forced transition*. Thus, the event mode is meaningful only for forced transitions. The firing mode sm is meaningful only for trigger transition.

Definition 2.1.4. (Marking, [Thi02])

Let $\mathcal{S}$ be an $_sNCE - structure$. A *marking* m of $\mathcal{S}$ is a mapping $m : P \rightarrow \{0, 1\}$.

$\square$

Furthermore, the modular and hierarchical characteristics are defined. The modular structure is constituted by the definition of an input/output set and the signal structure connecting the inputs and outputs to the $_sNCE - structure$.

Definition 2.1.5. (Input/Output- Set [Kar09])

An *Input/Output- set* (abbr.: I/O-set) is a tuple $\Phi = (C^{in}, E^{in}, C^{out}, E^{out})$ where

C^{in} is a finite set of *condition inputs*,

E^{in} is a finite set of *event inputs*,

C^{out} is a finite set of *condition outputs* and

E^{out} is a finite set of *event outputs*.

If every of those sets is empty, $\Phi = \emptyset$ is written. $\square$

Definition 2.1.6. (I/O- Structure)

Let $\mathcal{S}$ be an $_sNCE - structure$ and Φ an I/O-set. The *Input/Output-structure* (abbr.: I/O- structure) of $\mathcal{S}$ and Φ is a tuple $\Psi = (CN^{in}, EN^{in}, CN^{out}, EN^{out})$ with

$CN^{in} \subseteq (C^{in} \times T)$ is a finite set of *condition input arcs*,

$EN^{in} \subseteq (E^{in} \times T)$ is a finite set of *event input arcs*,

$CN^{out} \subseteq (P \times C^{out})$ is a finite set of *condition output arcs*, while
$$\forall c^{out} \in C^{out} : |\{p \in P | (p, c^{out}) \in CN^{out}\}| = 1,$$

$EN^{out} \subseteq (T \times E^{out})$ is a finite set of *event output arcs*, while
$$\forall e^{out} \in E^{out} : |\{t \in T | (t, e^{out}) \in EN^{out}\}| = 1.$$

If every of those sets is empty, $\Psi = \emptyset$ is written.

$\square$

The modularity of the model is obtained by encapsulating an $_sNCE - structure$ into module boundaries leading to *safe Net- Condition/Event Base Modules*. I/O's and the I/O-structure permit the interaction with an environment.

Definition 2.1.7. (safe Net- Condition/Event Base Module)

A *safe Net- Condition/Event Base Module* ($_sNM$) is a tuple

$$\mathcal{M}_B = (\mathcal{S}, m_0, \Phi, \Psi)$$

where
$\mathcal{S}$ is an $_sNCE - structure$,
$m_0 : P \rightarrow \{0, 1\}$ is an *initial marking* of $\mathcal{S}$,
Φ is an I/O- set and
Ψ is an I/O- structure.

$\square$

A module hierarchy is built based on the safe Net- Condition/Event Base Modules encapsulating the behavior model. Therefore, composite Net Condition/Event Modules are defined. Those can contain base modules and/or other composite modules.

The following definition of the hierarchical structure differs from the model hierarchy defined in [Thi02] and is based on the definition in [Kar09].

Definition 2.1.8. (safe Net- Condition/Event- Module)

A safe Net- Condition/Event Module ($_sNCEM$) is inductively defined as follows:

1. Every tuple $\mathcal{M}_B = (\mathcal{S}, m_0, \Phi, \Psi)$, which is an $_sNM$ is an $_sNCEM$ too.

2. If $\mathcal{M}_1, \mathcal{M}_2, ..., \mathcal{M}_k$ are $_sNCEM$, then is

$$\mathcal{M}_C = (\{\mathcal{M}_1, \mathcal{M}_2, ..., \mathcal{M}_k\}, \Phi, CK, EK)$$

an $_sNCEM$ too iff:

a) $\Phi = (C^{in}, E^{in}, C^{out}, E^{out})$ is an I/O-set.

b)

$$CK \subseteq \bigcup_{i \in \{1,...,k\}} (C^{in} \times C_i^{in}) \cup \bigcup_{i,j \in \{1,...,k\}} (C_i^{out} \times C_j^{in}) \cup \bigcup_{i \in \{1,...,k\}} (C_i^{out} \times C^{out})$$

describes the *condition interconnections* within $\mathcal{M}_C$, for which furthermore

$$\forall c_s \in \left(C^{out} \cup \bigcup_{i \in \{1,...,k\}} C_i^{in} \right) : |\{c_q \,|\, (c_q, c_s) \in CK\}| \le 1$$

applies.

c)

$$EK \subseteq \bigcup_{i \in \{1,...,k\}} (E^{in} \times E_i^{in}) \cup \bigcup_{i,j \in \{1,...,k\}} (E_i^{out} \times E_j^{in}) \cup \bigcup_{i \in \{1,...,k\}} (E_i^{out} \times E^{out})$$

describes the *event interconnections* within $\mathcal{M}_C$. It is supposed that

$$\forall e_S \in \left(E^{out} \cup \bigcup_{i \in \{1,...,k\}} E_i^{in} \right) : |\{e_q | (e_q, e_s) \in EK\}| \le 1.$$

$\mathcal{M}_C$ is called a *composite module*.

If all submoduls of $\mathcal{M}_C$ are base modules, than $\mathcal{M}_C$ is called *flat composite module*.

$Sub(\mathcal{M}_C) = \{\mathcal{M}_1, \mathcal{M}_2, \ldots, \mathcal{M}_k\}$ is a finite non empty set of safe Net- Condition/Event Modules. Every $\mathcal{M}_x \in Sub(\mathcal{M}_C)$ is called *submodule* of $\mathcal{M}_C$.

The *module set* $Mod(\mathcal{M})$ of an $(_sNCEM)$ is inductively defined as:

1. $\mathcal{M} \in Mod(\mathcal{M})$,

2. $\mathcal{M}' \in Mod(\mathcal{M})$ iff a composite module $\mathcal{M}^* \in Mod(\mathcal{M})$ exists, with $\mathcal{M}' \in Sub(\mathcal{M}^*)$.

$\square$

Composite modules do not include any net elements and therefore no own behavioral information. They provide the capability of hierarchical composition of models and can represent structural information of the modeled system.

The inductive nature of the Definition 2.1.8 ensures that no $_sNCEM$ contains itself. Thereby, cyclic relations of submodules are obviated.

Further, some notations are defined to simplify the following definitions. Preliminary to the successiv definition directed graphs are defined based on the the definition given in [Bra94].

Definition 2.1.9. *(Directed Graph)*
A *directed graph* G (or simply *graph*) is a tuple $G = (V, E)$ consisting of a non-empty *set of vertices* V and a set E of ordered pairs $E = \{(v, v') \,|\, v, v' \in V\}$ of vertices called arcs or directed edges or simply edges.

$\square$

Next, the linkage of transitions across module borders is specified.

Definition 2.1.10. (Event Chain, [Thi02])
Let $\mathcal{M}$ be an $_sNCEM$. The set of transitions within $\mathcal{M}$ is

$$\overline{T} := \bigcup_{\mathcal{M}_B \in Mod(\mathcal{M})} T_{\mathcal{M}_B}.$$

Further, be $G_E(\mathcal{M}) = (V_E(\mathcal{M}), E_E(\mathcal{M}))$ a *directed graph* with

$$V_E(\mathcal{M}) := \overline{T} \cup \bigcup_{M_x \in Mod(\mathcal{M})} \left(E^{in}_{\mathcal{M}_x} \cup E^{out}_{\mathcal{M}_x}\right)$$

and

$$E_E(\mathcal{M}) := \bigcup_{\mathcal{M}_C \in Mod(\mathcal{M})} EK_{\mathcal{M}_C} \cup \bigcup_{\mathcal{M}_B \in Mod(\mathcal{M})} \left(EN^{in}_{\mathcal{M}_B} \cup EN^{out}_{\mathcal{M}_B}\right).$$

An *event chain* (represented by $x \rightsquigarrow y$) exists from $x \in V_E(\mathcal{M})$ to $y \in V_E(\mathcal{M})$ iff

1. a *path* $x, \ldots y$ exists in $G_E(\mathcal{M})$, or

2.
$$(x, y) \in \bigcup_{\mathcal{M}_B \in Mod(\mathcal{M})} EN_{\mathcal{M}_B}.$$

Then x is called *event source* of y and y is called *event sink* of x, respectively.

Further it is required that the event chain $x \rightsquigarrow y$ is cycle free, i.e.

1. $\nexists (t_1, t_2) \in t_1 \rightsquigarrow t_2 : (t_1 = t_2)$ and

2. $\nexists ((t_1, t_2), \ldots, (t_{i-1}, t_i)) : (t_{l-1}, t_l) \in t_1 \rightsquigarrow t_2$ for $2 \leq l \leq i \wedge (t_1 = t_i)$.

□

An event chain between two transitions exists, if there is a closed sequence of event output arcs, event interconnections and event input arcs. A similar notation is introduced for condition signals.

Definition 2.1.11. (Condition Chain, [Thi02])
Let $\mathcal{M}$ be an $_sNCEM$. The set of transition within $\mathcal{M}$ is

$$\overline{P} := \bigcup_{\mathcal{M}_B \in Mod(\mathcal{M})} P_{\mathcal{M}_B}.$$

Further be $G_C(\mathcal{M}) = (V_C(\mathcal{M}), E_C(\mathcal{M}))$ a *directed graph* with

$$V_C(\mathcal{M}) := \overline{P} \cup \overline{T} \cup \bigcup_{M_x \in Mod(\mathcal{M})} \left(C_{\mathcal{M}_x}^{in} \cup C_{\mathcal{M}_x}^{out} \right)$$

and

$$E_C(\mathcal{M}) := \bigcup_{\mathcal{M}_C \in Mod(\mathcal{M})} CK_{\mathcal{M}_C} \cup \bigcup_{\mathcal{M}_B \in Mod(\mathcal{M})} \left(CN_{\mathcal{M}_B}^{in} \cup CN_{\mathcal{M}_B} \cup CN_{\mathcal{M}_B}^{out} \right).$$

A *condition chain* (represented by $x \rightarrow\!\!\bullet y$) exists from $x \in V_C(\mathcal{M})$ to $y \in V_C(\mathcal{M})$ iff a *path* $x, \ldots, y$ exists in $G_E(\mathcal{M})$. Then x is called *condition source* of y and y is called *condition sink* of x, respectively.

□

Not all of the inputs of a module are interconnected to other inputs or outputs in any case. Model modules in a library are often designed in a generalized way. That is to reduce the number of archived modules. Such modules can include elements excluded selectively or included as necessary. Some module inputs can be left unconnected to exclude the interconnected transitions. Signal sink transitions to unconnected inputs cannot fire because of the missing signals. That can be described in structural properties of transitions.

Definition 2.1.12. (Incompletely Controlled / Ordinary Transitions, [Thi02])
Be $\mathcal{M}$ an $_sNCEM$ with $\Phi = (C^{in}, E^{in}, C^{out}, E^{out})$.

$$\overline{C^{in}}(\mathcal{M}) := \bigcup_{M_x \in Mod(\mathcal{M}) - \{\mathcal{M}\}} C_{\mathcal{M}_x}^{in}, \quad \overline{E^{in}}(\mathcal{M}) := \bigcup_{M_x \in Mod(\mathcal{M}) - \{\mathcal{M}\}} E_{\mathcal{M}_x}^{in}$$

is *the set of enclosed condition-/ event- inputs of* $\mathcal{M}$.

A $c^{in} \in \overline{C^{in}}(\mathcal{M})$ is *connected* iff $\exists x \in C^{in} \cup \overline{P} : x \rightarrow\!\bullet c^{in}$. An $e^{in} \in \overline{E^{in}}(\mathcal{M})$ is *connected* iff $\exists x \in E^{in} \cup \overline{T} : x \rightsquigarrow e^{in}$.

$\overline{C_C^{in}}(\mathcal{M})$ / $\overline{E_C^{in}}(\mathcal{M})$ is the set of all connected condition/ event inputs of $\mathcal{M}$.

A transition $t \in \overline{T}$ is incompletely controlled iff

1. $\exists c^{in} \in \overline{C^{in}}(\mathcal{M}) - \overline{C^{in}}_C(\mathcal{M}) : c^{in} \rightarrow\!\bullet t$ or

2. $em(t) = \boxed{\wedge}$ and
   $\left(\exists e^{in} \in \overline{E^{in}}(\mathcal{M}) \text{ with } e^{in} \rightsquigarrow t : e^{in} \notin \overline{E_C^{in}}(\mathcal{M}) \right) \vee \left(\exists t' \in \overline{T} \text{ with } t' \rightsquigarrow t : t' \in \overline{T}_U \right)$ or

3. $em(t) = \boxed{\vee}$ and
   $\left(\forall e^{in} \in \overline{E^{in}}(\mathcal{M}) \text{ with } e^{in} \rightsquigarrow t : e^{in} \notin \overline{E_C^{in}}(\mathcal{M}) \right) \wedge \left(\forall t' \in \overline{T} \text{ with } t' \rightsquigarrow t : t' \in \overline{T}_U \right)$.

Then $\overline{T}_U$ is the *set of incompletely controlled transitions* of $\mathcal{M}$. $\overline{T}_G = \overline{T} - \overline{T}_U$ is the set of *ordinary transitions* of $\mathcal{M}$. Any transition $t \in \overline{T}$ is a ordinary transition within a base module $\mathcal{M}$ ($\overline{T}_G = \overline{T}$).

$\square$

The last main definition on the model syntax determines the top hierarchy level.

Definition 2.1.13. (safe Net- Condition/Event- System)

A *safe Net- Condition/Event- System* (abbr.: $sNCES$) is a tuple

$$\mathcal{M}_S = (\underline{\mathcal{M}}, CK, EK)$$

with

- $\underline{\mathcal{M}} = \{\mathcal{M}_1, \mathcal{M}_2, \ldots, \mathcal{M}_k\}$ a finite, non empty set of $sNCEM$,

-

$$CK \subseteq \bigcup_{i,j \in \{1,2,\ldots,k\}} (C_i^{out} \times C_j^{in})$$

describing the *condition interconnections* in $\mathcal{M}_S$, while further holds

$$\forall c_s \in \bigcup_{j \in \{1,2,\ldots,k\}} C_j^{in} : |\{c_q|(c_q, c_s)\}| \leq 1,$$

-

$$EK \subseteq \bigcup_{i,j \in \{1,2,\ldots,k\}} \left(E_i^{out} \times E_j^{in} \right)$$

describing the *event interconnections* in $\mathcal{M}_S$. It is required that

$$\forall e_s \in \bigcup_{j \in \{1,2,\ldots,k\}} E_j^{in} : |\{e_q|(e_q, e_s) \in EK\}| \leq 1.$$

If all modules of $\mathcal{M}_S$ are base modules, than $\mathcal{M}_S$ is called *base system*. The module set $Mod(\mathcal{M}_S)$ of an $_sNCES$ is defined as

$$Mod\,(\mathcal{M}_S) := \bigcup_{i \in \{1,2,\ldots,k\}} Mod\,(\mathcal{M}_i) \text{ with } \mathcal{M}_i \in \underline{\mathcal{M}}$$

□

A safe Net- Condition/Event- System consists of a network of modules. A system doesn't contain any inputs or outputs in contrast to the definition in [Thi02]. Therefore, the $_sNCES$ describes the highest non-composable hierarchy level of a model. Naturally, that has to be reflected in the model semantic.

Finally, the notations for signal chains are defined to hold for $_sNCES$ too.

Definition 2.1.14. (Signal Chains in safe Net- Condition/Event- Systems)
Be $\mathcal{M}_S$ a safe Net- Condition/Event- System, then Definition 2.1.10, 2.1.11 and 2.1.12 are applied to $\mathcal{M}_S$ through simple replacing $\mathcal{M}$ by $\mathcal{M}_S$.

□

2.2 Modular semantic

The semantic of $NCES$ [Kar09] is defined in terms of transition sets called steps. The steps are specified by the interconnections via event signals and the event mode of transitions. Only enabled steps cause a state transition. The semantic is described differing between modules and the system corresponding to the structure definitions. Thus, the analysis of model dynamics is provided on the module level and without the need of composition of the whole system. That supports especially the analysis of distributed system models. The elements of a modular system can be analyzed separately or by modular algorithms. The following semantic definitions are orientated on the definition in [Thi02]. They are modified in a way to meet the reviewed model syntax and split into a local and a (sub)- system view. This allows behavioral analysis of models or model parts without previous complete composition [MH07].

To allow modular and cross- modular step definitions, the input state of modules is defined in principle and under consideration of a concrete step.

The input state of $_sNCE$ Modules is defined as follows:

Definition 2.2.1. (Input State)

The input state is of an $_sNCEM$ is a mapping $is : C^{in} \cup E^{in} \rightarrow \{0,1\}$ assigning a value of $\{0,1\}$ to each signal input.

$\square$

Concrete input states are defined under consideration of steps defined in Definition 2.2.8.

Definition 2.2.2. (Concrete Input State)

The input state is of a module $\mathcal{M}_i$ in a composite module $\mathcal{M}$ or $_sNCES$ $\mathcal{M}_S$ results from the values of signal inputs $c_i^{in} \in C_{\mathcal{M}_i}^{in}$ and $e_i^{in} \in E_{\mathcal{M}_i}^{in}$ under consideration of a step ξ as follows:

$$c_i^{in} = \begin{cases} 1 & if \; \exists p \in \overline{P} : p \rightarrow\!\bullet c_i^{in} \wedge m(p) = 1 \\ 0 & else, \end{cases}$$

$$e_i^{in} = \begin{cases} 1 & if \; \exists t \in \xi : t \rightsquigarrow e_i^{in} \\ 0 & else. \end{cases}$$

$\square$

The values of signal inputs depend on a known environment of the module. The internal behavior of a module in an unknown environment is analyzed under consideration of all possible input states. Thereby the derived information is true for the assumed input state only.

In $_sNCES$, no input states have to be assumed. The signal couplings are expected to be complete. The input state of unconnected inputs is zero following Definition 2.2.2.

The enabling of a transition and hence of a step within a module depends on the marking and the input state of a module only. The marking influences the enabling through ordinary arcs, similar to Petri nets and additionally through condition arcs.

Hence, two types of conditions are defined for the enabling of transitions.

Definition 2.2.3. (Enabling of Transitions)

A transition $t \in T$ of an $_sNCEM$ is:

1. *Marking enabled* at a marking m iff
   $(\forall p \in P \; with \; (p,t) \in F : m(p) = 1) \wedge (\forall p \in P \; with \; (t,p) \in F : m(p) = 0),$

2. *Condition enabled* at a marking m and an input state is iff
   $\forall p \in P \; with \; (p,t) \in CN : m(p) = 1$ and
   $\forall c^{in} \in C^{in} \; with \; (c^{in},t) \in CN^{in} : is(c^{in}) = 1.$

$\square$

A transition is marking enabled if all pre-places are marked and all post-places do not hold a token. It is condition enabled if all source places of associated condition arcs or condition signal interconnections are marked.

The marking of the source place of a signal interconnection is represented by the local input state following Definition 2.2.1.

Marking and condition enabled transitions forming a step have to be free of conflicts as used in terms of Petri nets.

Definition 2.2.4. (Conflict, Contact)
Let $\mathcal{M}$ be an $_sNCEM$, m a marking and $\xi \subseteq T$ a set of transitions of $\mathcal{M}$ which are marking and condition enabled in m.

ξ is under *conflict* in m iff $\exists p \in {}^F\xi : \left|p^F \cap \xi\right| > 1$.

ξ is under *contact* in m iff $\exists p \in \xi^F : \left|{}^F p \cap \xi\right| > 1$.

$\square$

Based on these terms, the definition of *modular steps* in general and *enabled modular steps* under a concrete marking are introduced. The event mode and the firing mode are considered in particular. The steps are called modular because they depend on the analyzed base module and its input state only.

Definition 2.2.5. (Modular Steps of Base Modules)
Let $\mathcal{M}_B$ be an $_sNCEM$ with the marking m, the input state *is* and $\xi_{\mathcal{M}} \subset T$ a non empty set of transitions within a module $\mathcal{M}_B$.
$\xi_{\mathcal{M}}$ is a *modular step* within the module $\mathcal{M}_B$ iff

1. $|\xi_{\mathcal{M}} \cap (T^t)| \geq 1 \dot{\vee} |\xi_{\mathcal{M}} \cap (T_{Ei} \cup T_{Es})| = 1$,
   while $T^t := \{t^t \in T | \exists e^{in} : (e^{in}, t) \in EN^{in}\}$,

2. for every transition $t \in \xi_{\mathcal{M}} : t \notin (T_{Ei} \cup T_{Es})$ holds:

   - $em(t) = \boxtimes \wedge ((\exists t^{'} \in \xi_{\mathcal{M}} : (t^{'}, t) \in EN) \vee (\exists e^{in} \in E^{in} \text{ with } (e^{in}, t) \in EN^{in} : is(e^{in}) = 1))$ or

   - $em(t) = \boxtimes \wedge ((\forall t^{'} \text{ with } (t^{'}, t) \in EN : t^{'} \in \xi_{\mathcal{M}}) \wedge (\forall e^{in} \in E^{in} \text{ with } (e^{in}, t) \in EN^{in} : is(e^{in}) = 1))$
     and

3. all transitions are free of conflicts and contacts to each other.

$\Xi_{\mathcal{M}}$ is the set of steps within $\mathcal{M}$.

$\xi_{\mathcal{M}}$ is called *enabled modular step* under m and *is* iff

- $\forall t \in xi_{\mathcal{M}}$ holds they are marking and condition enabled under m and *is* and there is no set of transitions with $\xi'_{\mathcal{M}} = \xi_{\mathcal{M}} \cup \{t\}$ within $\mathcal{M}$, which is also a step and marking and condition enabled under m and *is* and $\left| \xi_{\mathcal{M}} \cap \overline{T_{Ei}} \right| = 1$, or

- $\forall t \in \xi_{\mathcal{M}}$ holds they are marking and condition enabled under m and *is* and there is no set of transitions with $\xi'_{\mathcal{M}} = \xi_{\mathcal{M}} \cup \{t\}$ within $\mathcal{M}$, which is also a step and marking and condition enabled under m and *is* and no enabled modular step exists with $\left| \xi_{\mathcal{M}} \cap \overline{T_{Ei}} \right| = 1$.

$\square$

Modular steps can be determined depending on an assumed or known input state (considering a known environment). A modular step contains either one *trigger transition* t^t or any number of transitions triggered externally by activated event inputs called *local trigger transitions*. For local trigger transitions the property of being *event input enabled* is defined.

Definition 2.2.6. (Event Input Enabling)
A local trigger transition is event input enabled if condition 2) of Def.2.2.5 is satisfied.
$\square$

Figure 2.2 shows an example of an $_sNCE$ Module. The local step $\xi_{M_4} = (t1, t2, t6)$ is enabled under the shown marking if the event inputs $ei1$ and $ei2$ are true ($is(ei1) = is(ei2) = 1$). Event input arcs lead from these inputs to $t2$ and $t6$ and identify them as local trigger transitions.

Transitions $t^t_{\mathcal{M}}$ are called *local trigger transitions* of $\mathcal{M}$.

Steps in composite modules are defined as aggregation of local steps.

Definition 2.2.7. (Steps in an $_sNCEM$)
Let $\mathcal{M}$ be an $_sNCEM$ with the marking m, the input state *is* and $\xi \in \overline{T}$ a nonempty set of transitions within $\mathcal{M}$.
A *step* ξ within $\mathcal{M}$ is defined as follows:

1. A local step $\xi_{\mathcal{M}}$ is a step ξ if $\left| \xi \cap (\overline{T_{Ei}} \cup \overline{T_{Es}}) \right| = 1$ holds.

2. Every union of local steps

$$\xi = \bigcup_{M_x \in Mod(\mathcal{M})} \xi_{M_x} \text{ is a step if}$$

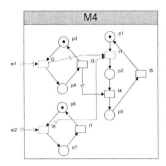

Figure 2.2: Example Net Condition/Event Module with the enabled local step $\xi_{M_4} = (t1, t2, t6)$.

- $|\xi \cap (\overline{T_{Ei}} \cup \overline{T_{Es}})| = 1$ holds and
- for every transition $t^t \in \xi_{M_x}$ holds:
  - $em(t^t) = \boxvoid \wedge (\exists t' \in \xi | t' \rightsquigarrow t^t)$ or
  - $em(t^t) = \boxtimes \wedge (\forall t' \in \overline{T} \text{ with } t' \rightsquigarrow t^t : t' \in \xi)$.

Ξ is the set of steps within $\mathcal{M}_S$.

ξ is called *enabled step* under m and *is* iff

- all $\xi_{\mathcal{M}} \subseteq \xi$ are enabled under m and *is* and there is no set of local steps with $\xi' = \xi \cup \{\xi_{\mathcal{M}}\}$, which is also enabled under m and *is* and further holds $|\xi \cap (\overline{T_{Ei}})| = 1$ or

- all $\xi_{\mathcal{M}} \subseteq \xi$ are enabled under m and *is* and there is no set of local steps with $\xi' = \xi \cup \{\xi_{\mathcal{M}}\}$, which is also enabled under m and *is* and no enabled step exists with $|\xi \cap (\overline{T_{Ei}})| = 1$.

□

The defined enabled steps are always maximal steps and contain exactly one trigger transition, i.e. there is no marking and condition enabled transition, whose inclusion would be an enabled step too. Conflict transitions must not be part of the same step.

In Figure 2.3 a modular step and the included local steps are shown schematically. The example of a modular step includes the local step displayed in Figure 2.2. The step can be seen as a network of interconnected local steps and is triggered by exactly one trigger transition.

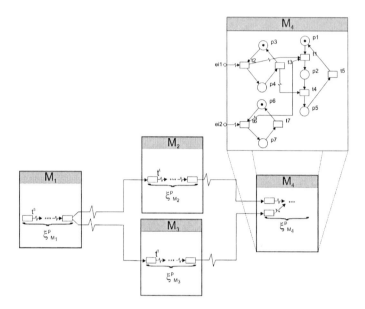

Figure 2.3: Schematic illustration of a modular step in $_sNCES$.

Definition 2.2.8. (Steps in an $_sNCES$)
Let M_S be a safe Net- Condition/Event- System, then Definition 2.2.7 and the following Definitions 2.2.9, 2.2.10 are applied to $\mathcal{M}_S$ through simply replacing $\mathcal{M}$ by $\mathcal{M}_S$. $\square$

The effect of firing an enabled step is defined as follows.

Definition 2.2.9. (Successor Marking)
Let $\mathcal{M}$ be an $_sNCEM$ with the marking m and the input state is.
If ξ is an enabled step under m and is, then ξ is enabled to fire. The successor marking m' is determined for $p \in \overline{P}$ to

$$
m'(p) = \begin{cases} 1 & if\ \exists t \in \xi : (t,p) \in F \\ 0 & if\ \exists t \in \xi : (p,t) \in F \\ m(p) & else. \end{cases}
$$

The notation $m(\mathcal{M})\,[\xi\rangle\,m'(\mathcal{M})$ means that $m'(\mathcal{M})$ is the successor marking of $m(\mathcal{M})$ by firing the enabled step ξ. $\square$

Sequences of realized steps and their successor markings, starting from an initial marking, form the set of reachable markings.

Definition 2.2.10. (Firing Sequence / Reachable States)

Be $\mathcal{M}$ an $_sNCEM$. A marking m'' of $\mathcal{M}$ is called *follower marking* of m_0 at w $(m_0\,[w\rangle\,m'')$, while $w = \xi_1, \ldots, \xi_n \in \Xi$ iff

1. $|w| = 0 \wedge m_0 = m''$ holds, or

2. there exists a marking m' with

$$m_0\,[(\xi_1 \ldots \xi_{n-1})\rangle\,m' \wedge m'\,[\xi_n\rangle\,m''.$$

The w is called *firing sequence* in that case. Further, w is called activated under m_0 and $m_0\,[w\rangle$ is written.

$$\mathcal{W}_{\mathcal{M}} := \{w \in \xi | m_0\,[w\rangle\}$$

is the *set of all feasible firing sequences* in $\mathcal{M}$.

$$[m_0\rangle_{\mathcal{M}} := \{\exists w \in \mathcal{W}_{\mathcal{M}} : m_0\,[w\rangle\,m\}$$

is the *set of reachable markings*. The set of executable steps $AS_{\mathcal{M}}$ of $\mathcal{M}$ is

$$AS_{\mathcal{M}} = \{\xi \in \Xi | \exists m \in [m_0\rangle_{\mathcal{M}} : m\,[\xi\rangle\}.$$

A step ξ with $\xi \in AS_{\mathcal{M}}$ is an executable step. A sequence $[m_0, \xi_1, m_1, \xi_2, \ldots, \xi_k, m_k]$ with

1. $\xi_1 \ldots \xi_k \in \mathcal{W}_{\mathcal{M}}$ and

2. $\forall 1 \leq i \leq k : m_{i-1}\,[\xi_i\rangle\,m_i$

is called trajectory in $\mathcal{M}$. $\mathcal{TR}_{\mathcal{M}}$ is the set of all trajectories.

$\square$

The terms reachable marking of a model and reachable state are used synonymous in this work. The set of reachable states and the set of firing sequences form the explicit and complete image of the behavior modeled implicitly by the given $_sNCES$.

2.3 Composition

Composition is generally used to join independently created model components. They can model functional varying components as plant and controller or disaggregated elements of a system. It is necessary to compose the components to analyze the system behavior as a whole. The process of assembling and interconnecting model components according to a formal rule is called composition in this work.

The term composition describes the merging of components (e.g. base modules) to a new model entity. It is introduced in [Thi02] and just shortly described in the following. The two defined composition types are transformation rules on the modular and hierarchical structure of a model. First, the vertical composition transforms a hierarchical $_sNCEM$ to a flat composite module (according to Definition 2.1.8).

Definition 2.3.1. (Vertical Composition - Module, [Thi02])
Let $\mathcal{M} = (Sub\,(\mathcal{M}), \Phi, CK, EK)$ be a $_sNCEM$. *A flat composite module*
$\mathscr{C}_{\mathcal{V}}(\mathcal{M}) = (M_{\mathscr{C}}, \Phi, CK_{\mathscr{C}}, EK_{\mathscr{C}})$ *is composed vertically from $\mathcal{M}$ iff*

1. $M_{\mathscr{C}} = \{\mathcal{M}_B | \mathcal{M}_B \in Mod\,(\mathcal{M})\}$,

2.

$$CK_{\mathscr{C}} = \begin{cases} \left(c^{in}, c_1^{in}\right) | c^{in} \in C^{in} \wedge \exists \mathcal{M}_{B/1} \in M_{\mathscr{C}} : c_1^{in} \in C_{B/1}^{in} \wedge c^{in} \relbar\joinrel\bullet c_1^{in} \end{cases} \cup \\ \left(c_1^{out}, c_2^{in}\right) | \exists \mathcal{M}_{B/1}, \mathcal{M}_{B/2} \in M_{\mathscr{C}} : c_1^{out} \in C_{B/1}^{out} \wedge c_2^{in} \in C_{B/2}^{in} \wedge c_1^{out} \relbar\joinrel\bullet c_2^{in} \Big\} \cup \\ \left(c_1^{out}, c^{out}\right) | c^{out} \in C^{out} \wedge \exists \mathcal{M}_{B/1} \in M_{\mathscr{C}} : c_1^{out} \in C_{B/1}^{out} \wedge c_1^{out} \relbar\joinrel\bullet c^{out} \Big\},$$

3.

$$EK_{\mathscr{C}} = \begin{cases} \left(e^{in}, e_1^{in}\right) | e^{in} \in E^{in} \wedge \exists \mathcal{M}_{B/1} \in M_{\mathscr{C}} : e_1^{in} \in E_{B/1}^{in} \wedge e^{in} \rightsquigarrow e_1^{in} \end{cases} \cup \\ \left(e_1^{out}, e_2^{in}\right) | \exists \mathcal{M}_{B/1}, \mathcal{M}_{B/2} \in M_{\mathscr{C}} : e_1^{out} \in E_{B/1}^{out} \wedge e_2^{in} \in E_{B/2}^{in} \wedge e_1^{out} \rightsquigarrow e_2^{in} \Big\} \cup \\ \left(e_1^{out}, e^{out}\right) | e^{out} \in E^{out} \wedge \exists \mathcal{M}_{B/1} \in M_{\mathscr{C}} : e_1^{out} \in E_{B/1}^{out} \wedge e_1^{out} \rightsquigarrow e^{out} \Big\}.$$

$\square$

The result of the vertical composition is a flat composite module containing a network of base modules. The base modules themselves remain unchanged. The vertical composition dissipates the hierarchy level between the composed module and the base modules. Figure 2.4 exemplifies the modification of the hierarchical structure by vertical composition of a composite module.

The following definitions of the horizontal composition are transfused from the definition on $TNCES$ in [Thi02] to $_sNCES$.

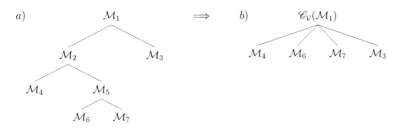

Figure 2.4: Example of the vertical composition of modules with multiple intermediate levels.

Definition 2.3.2. (Horizontal Composition - $_sNCE$-Structure)
Let $\mathcal{M} = (Sub(\mathcal{M}), \Phi, CK, EK)$ be a flat composite module with
$Sub(\mathcal{M}) = \{\mathcal{M}_1, \mathcal{M}_2, \ldots, \mathcal{M}_x\}$.
The $_sNCE$-structure $\mathcal{S}_{\mathscr{C}}(\mathcal{M}) = (\mathcal{S})$ is *composed horizontally* of $\mathcal{M}$ iff
$\mathcal{S} = (P, T, F, CN, EN, em, sm)$, while

- $$P = \overline{(P)},\ T = \overline{T_G},\ F = \bigcup_{i \in \{1,\ldots,x\}} \left\{ f \in F_i | f \in \left(\overline{P} \times \overline{T_G} \cup \overline{T_G} \times \overline{P}\right) \right\},$$

- $CN = \{(p,t) \, | \, p \in P \wedge t \in T \wedge p \rightarrow\!\!\bullet\, t\}$,

- $EN = \{(t_1, t_2) \, | \, t_1, t_2 \in T \wedge t_1 \rightsquigarrow t_2\}$,

- $\forall t \in T : em(t) = em_i(t)$, in case $t \in T_i$ with $i \in \{1, \ldots, x\}$,

- $\forall t \in T : sm(t) = sm_i(t)$, in case $t \in T_i$ with $i \in \{1, \ldots, x\}$.

$\square$

Constitutively, the horizontal composed module can be defined by specification of the I/O-structure and the initial state.

Definition 2.3.3. (Horizontal Composition - Module)
Let $\mathcal{M} = (Sub(\mathcal{M}), \Phi, CK, EK)$ be a flat composite module with
$Sub(\mathcal{M}) = \{\mathcal{M}_1, \mathcal{M}_2, \ldots, \mathcal{M}_x\}$ and $\Phi = (C^{in}, E^{in}, C^{out}, E^{out})$. Further be $\mathcal{S}_{\mathscr{C}}(\mathcal{M})$ a
$_sNCE$-structure with $\mathcal{S}_{\mathscr{C}} = (P, T, F, CN, EN, em, sm)$ horizontally composed of $\mathcal{M}$.
The $_sNM\ \mathscr{C}_{\mathcal{H}}(\mathcal{M}) = (\mathcal{S}_{\mathscr{C}}(\mathcal{M}), m_0, \Phi, \Psi(\mathcal{S}_{\mathscr{C}}(\mathcal{M}), \Phi))$ is *horizontally composed* of $\mathcal{M}$ iff

1. $\forall p \in P : m_0(p) = m_{0i}(p)$, in case $p \in P_i$ with $i \in \{1, \ldots, x\}$ and

2. $\Psi\left(\mathcal{S}_{\mathscr{C}}(\mathcal{M}), \Phi\right) = (CN^{in}, EN^{in}, CN^{out}, EN^{out})$, while

- $CN^{in} = \{(c^{in}, t) \mid c^{in} \in C^{in} \wedge t \in T \wedge c^{in} \rightarrow\!\bullet t\}$,

- $EN^{in} = \{(e^{in}, t) \mid e^{in} \in E^{in} \wedge t \in T \wedge e^{in} \rightsquigarrow t\}$,

- $CN^{out} = \{(p, c^{out}) \mid p \in P \wedge c^{out} \in C^{out} \wedge p \rightarrow\!\bullet c^{out}\}$,

- $EN^{out} = \{(t, e^{out}) \mid t \in T \wedge e^{out} \in E^{out} \wedge t \rightsquigarrow e^{out}\}$.

$\square$

The result of the horizontal composition of a module $\mathcal{M}$ is a base module $\mathscr{C}_{\mathcal{H}}(\mathcal{M})$ with the same interface as $\mathcal{M}$. The internal structure of the composed module contains all the same signal interconnections as the original composite module. Module borders and the I/O- structures of the contained base modules are dissolved by horizontal composition. Thereby, the incompletely controlled transitions are eliminated from the net structure $(\overline{T} - \overline{T_U} \Rightarrow \overline{T})$. They would convert to ordinary trigger transition by deleting the associated input structure, what is not permitted. Hence, the horizontally composed image of $\mathcal{M}$ contains only ordinary transitions. All controllable transitions supposed to be part of the composed module must have an event connection to an input of the supreme module (the composed one). So they are ordinary according to Definition 2.1.12.

The transformation of a flat composite module to a base module through horizontal composition is illustrated in Figure 2.5 with the vertically composed module $\mathcal{M}_1$ shown in Figure 2.4.

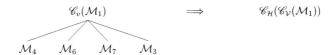

Figure 2.5: Result of the horizontal composition of the vertically composed module in Figure 2.4.

$_sNCE$ Systems are not covered by the definitions introduced so far. The composition rules defined in Definition 2.3.1-2.3.3 cannot be applied to the systems as defined in Definition 2.1.13, but the order of transformation steps is the same as in the composition of modules. A vertical composition is applied to all modules of the system. The resulting network of base modules is composed horizontally, thereafter.

Definition 2.3.4. (Vertical Composition of a System)
Let $\mathcal{M}_S = (\underline{\mathcal{M}}, CK, EK)$ be an $_sNCES$ with $\underline{\mathcal{M}} = \{\mathcal{M}_1, \mathcal{M}_2, \ldots, \mathcal{M}_x\}$.
The base system $\mathscr{C}_{V,S} = (\underline{\mathcal{M}_\mathscr{C}}, CK, EK)$ with $\underline{\mathcal{M}_\mathscr{C}} = \{\mathcal{M}_{\mathscr{C}1}, \mathcal{M}_{\mathscr{C}2}, \ldots, \mathcal{M}_{\mathscr{C}i}\}$ is *vertically composed* of $\mathcal{M}_S$ iff

$$\forall \mathcal{M}_{\mathscr{C},i} \in \underline{\mathcal{M}_\mathscr{C}} : \mathcal{M}_{\mathscr{C},i} = \mathscr{C}_\mathcal{H}\left(\mathscr{C}_\mathcal{V}\left(\mathcal{M}_i\right)\right), \, i \in \{1, 2, \ldots, c\},$$

while

$$\mathcal{M}_i \in Mod(\underline{\mathcal{M}}) \wedge \not\exists \mathcal{M}_i^* \in Mod(\underline{\mathcal{M}}) : \mathcal{M}_i \in Sub(\mathcal{M}_i^*).$$

$\square$

The resulting base system is converted into a base module in two steps. First the composed $_sNCE$- structure of the module is defined.

Definition 2.3.5. (Horizontal Composition of a System - $_sNCE$-Structure)
Let $\mathcal{M}_S = (\underline{\mathcal{M}}, CK, EK)$ be a base system with $\underline{\mathcal{M}} = \{\mathcal{M}_1, \mathcal{M}_2, \ldots, \mathcal{M}_x\}$.
The $_sNCE$-structure $\mathcal{S}_\mathscr{C}(\mathcal{M}_S) = (\mathcal{S})$ is *horizontally composed* of $\mathcal{M}_S$ iff

- $$P = \overline{(P)}, \, T = \overline{T_G}, \, F = \bigcup_{i \in \{1, \ldots, x\}} \left\{ f \in F_i | f \in \left(\overline{P} \times \overline{T_G} \cup \overline{T_G} \times \overline{P}\right)\right\},$$

- $CN = \{(p, t) \,| p \in P \wedge t \in T \wedge p \multimap t\}$,

- $EN = \{(t_1, t_2) \,| t_1, t_2 \in T \wedge t_1 \leadsto t_2\}$,

- $\forall t \in T : em(t) = em_i(t)$, in case $t \in T_i$ with $i \in \{1, \ldots, x\}$,

- $\forall t \in T : sm(t) = sm_i(t)$, in case $t \in T_i$ with $i \in \{1, \ldots, x\}$.

$\square$

Finally, a module enclosure for the $_sNCE$-structure and its initial marking are specified.

Definition 2.3.6. (Horizontal Composition of a System - Module)
Let $\mathcal{M}_S = (\underline{\mathcal{M}}, CK, EK)$ be a base system with $\underline{\mathcal{M}} = \{\mathcal{M}_1, \mathcal{M}_2, \ldots, \mathcal{M}_x\}$. Further, be $\mathcal{S}_\mathscr{C}(\mathcal{M}_S)$ a horizontally composed $_sNCE$- structure of $\mathcal{M}_S$.
The $_sNM$ $\mathscr{C}_\mathcal{H}(\mathcal{M}_S) = (\mathcal{S}_\mathscr{C}(\mathcal{M}_S), m_0)$ is horizontally composed of $\mathcal{M}_S$ iff
$\forall p \in P : m_0(p) = m_{0i}(p)$ in case $p \in P_i$ with $i \in \{1, \ldots, x\}$.

$\square$

The resulting module must not include an I/O- set and no I/O- structure because $_sNCE$-Systems do not contain any interface. The signal interconnections are transformed to signal arcs directly connecting the net nodes. The composition to a new module is required to generate an image representing the model behavior of the original $_sNCE$- System. The vertical and horizontal compositions of systems are applicable separately with respect to the intended analysis, similar to modules. The analysis of closed-loop models is processed only on fully composed models until now. The *TNCES-workbench* [TNCES] includes an automatic full composition of the model to be analyzed.

2.4 Structural analysis

Structural properties of a model or of model parts are used in most model-based algorithms. Structural properties known from *Petri nets* are place and transition invariants (p- and t-invariants) [DR98]. They allow some predications on the model behavior without explicit construction of all reachable markings. The basis of structural analysis methods on Net Condition/Event Systems is much lower than in the commonly used models as Petri nets. However, there exists some work on determination of invariants of $NCES$. The potential of using invariants of the Petri net parts (*underlying Petri nets*) of an $NCES$ is discussed in [SR02].

Beside the p- invariants of the underlying Petri nets, a new structural property called *event synchronization* is defined for $_sNCES$ in this section. The event synchronization property will be used especially in algorithms using partial markings.

2.4.1 State invariants

A *state invariant* $\mathcal{I}$ (*S*-invariant for short) is a non-constant mapping defined on the set of all imaginable states, which is constant on the set of all reachable states [SR02]. Whether $\mathcal{I}$ is a state invariant of the considered system or not depends on the initial marking, just as any statement referencing to reachability (Definition 2.2.10). For $_sNCES$ as for Petri nets, the set of all imaginable states is the set of all markings M_i.

Theorem 1. *[SR02]*
Every (linear) state invariant of the underlying Petri net PN is a (linear) state invariant of $\mathcal{M}$ or $\mathcal{M}_S$, respectively.

That is because the marking flow of $\mathcal{M}$ or $\mathcal{M}_S$ is realized exclusively by the elements (nodes and ordinary arcs) of the underlying Petri net.

A determination of state invariants in general is described in [SR02] based on the set of executable steps under a given initial marking. But the set of all executable steps can only be identified under knowledge of the set of all reachable states. Therefore, this approach does not have meaning in analysis. In the following place invariants and event synchronizations are introduced for $_sNCES$. Both are state invariants and determinable without reachable state space computation.

2.4.2 Place invariants

The place invariant property is commonly used in analysis of Petri nets and its derivates. A place invariant of a Petri net specifies an integer valued weighting function i on a non empty finite set of places for which the weighted sum of tokens stays constant under marking flow.

$$\sum_{p \in {}^F t} i(p) = \sum_{p \in t^F} i(p)$$

Consequently it has to hold $i \circ m = i \circ m_0$ for all reachable markings of $\mathcal{M}/\mathcal{M}_S$. In other words, the place invariant of an $_sNCES$ specifies a set of places for which the sum of tokens stays constant for all reachable markings. The check of place invariants allows a simple non-reachability test.

In difference to Petri nets in $_sNCES$ not only single transitions have to be considered but the firing of steps ξ. Hence the specification of place invariants is modified to

Theorem 2. *[SR02]*
For any $_sNCE-$ Module / System $\mathcal{M}/\mathcal{M}_S$ holds:

1. *A non-zero P-vector i is a place invariant of $\mathcal{M}/\mathcal{M}_S$ iff $i \circ C_{\mathcal{M}_{(S)}} = 0$, while $C_{\mathcal{M}_{(S)}}$ denote the matrix with rows corresponding to the places and columns formed by the P-vectors $\Delta\xi$ ($\Delta\xi = \xi^+ - \xi^-$) for $\xi \in \Xi$.*

2. *Any P-invariant of PN is a P-invariant of $\mathcal{M}/\mathcal{M}_S$.*

C_M denotes the matrix with rows corresponding to the places and columns formed by the place-vector $\Delta\xi$ for $\xi \in \Xi$ the set of enabled steps. While the first assertion again refers to the set of executable steps Ξ, the second one leads to a common way of P-invariant evaluation. It follows from the fact that the columns of $C_{\mathcal{M}_{(S)}}$ are sums of columns of

the incidence matrix of the underlying Petri net. That is because steps are transition
sets and the marking flow through steps is the sum of the flow of the contained single
transitions. The determination of p-invariants of Petri nets is described for example in
[DR98, MS82, Sta90].

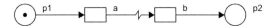

Figure 2.6: Counter example to the converse of Theorem 2.

The converse of the second assertion of Theorem 2 is not true. Figure 2.6 presents a counter
example. The only enabled (executable) step is $\xi = \{a, b\}$. A place invariant of $\mathcal{M}$ is given
by the place vector $i = \begin{pmatrix} 1 \\ 1 \end{pmatrix}$. Obviously, there exists no p-invariant for the underlying
Petri net.

P- invariants (in Petri nets and $_sNCES$) are independent of an initial marking. The
example net in Figure 2.7 contains the two place invariants $i = \{1, 1, 1, 0, 0\}$ and $i =
\{0, 0, 0, 1, 1\}$.

2.4.3 Event synchronization

Unlike p- and t- invariants based on the underlying Petri net, the *event synchronizations*
are structural properties explicitly based on signals, more precisely event interconnections.
An event signal synchronizes two transitions in one direction under the enabling conditions.
That means that the event source forces the event sink to synchronous firing if both are
enabled. The event sink cannot fire on its own or force the signal source. A set of pairs of
transitions interconnected by event arcs can lead to state invariants not covered by place
invariants. An example is shown in Figure 2.7.

The reachable states are $(p3, p4)$, $(p2, p5)$ and $(p1, p5)$. The example contains the two p-
invariants $m(p1) + m(p2) + m(p3) = 1$ and $m(p4) + m(p5) = 1$. Further it can be seen
that $m(p3) + m(p5) = 1$ holds always true under the given initial marking. The subnet
with $p4, p5$ is fully synchronized by event interconnections. The place influenced ($p5$) by
the event sinks is called to be *event-synchronized* with the place $p3$ surrounded by the
source transitions. The additional state invariant $m(p1) + m(p2) + m(p4) = 1$ can be
constructed by the combination of p- invariants and event synchronization. The simple
example situation is generalized to places synchronized to a set of places in the following.

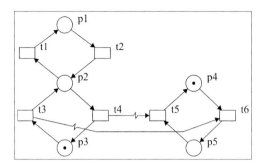

Figure 2.7: Example net for event-synchronized places ($p3, p4$).

A place p is synchronized with a set of places P^* if all transitions of the pre-region and of the post-region of p and the set P^* are interconnected by event arcs and there are no additional condition arcs at the pre- and post-transitions of p. The event interconnections have to be directed from $t \in {}^F P^* \cup P^{*F}$ to all transitions surrounding p. The token flow from and to p has to be influenced only by the forced transitions. The synchronization denotes that the synchronized place isn't marked if one of the places of P^* holds a token.

Following the condition of being event-synchronized is defined formally:

Definition 2.4.1. Event Synchronization
Let $\mathcal{M}/\mathcal{M}_S$ be an $_sNCES/_sNCEM$ with the initial marking m_0. A place p is *event-synchronized* with a set of places $p^* \in P^*$ if the following holds:

- $\forall t \in {}^F p | \exists (t', t) \in EN \land \forall t' \in T : \left| {}^F t' \cap P^* \right| = 1$,

- $\forall t' \in p^{*F} | \exists t \in T : (t', t) \in EN \land t \in {}^F p$,

- $\forall t \in p^F | \exists (t'', t) \in EN \land \forall t'' \in T : \left| t''^F \cap P^* \right| = 1$,

- $\forall t'' \in {}^F p^* | \exists t \in T \land (t'', t) \in EN : t \in p^F$ and

- ${}^F \left({}^F p \right) = \left(p^F \right)^F \land \left({}^F p \right)^F = p \land {}^F \left(\left(p^F \right)^F \right) = p^F \land {}^F \left({}^F \left({}^F p \right) \right)^F = {}^F p \land {}^{CN} p \cup p^{CN} = \emptyset$.

For the set of places $p \cup P^*$ with an event synchronization of p to the set $P^* = \{p_1^*, \ldots, p_k^*\}$ holds:

$$m(p) + m(p_1^*) + \ldots + m(p_k^*) = 1, \text{ if it holds for } m_0.$$

□

The determination of event synchronization is described by the pseudo code Algorithm 1. It returns the state invariants.

foreach $p \in P$ **do**
   if *all* $t \in \{^F p \cup p^F\}$ *are event trigger and for p holds*
   $^F\left(^F p\right) = \left(p^F\right)^F \wedge \left(^F p\right)^F = p \wedge \left(\left(p^F\right)^F\right)^F = p^F \wedge \left(^F\left(^F p\right)\right)^F = {}^F p \wedge {}^{CN} p \cup p^{CN} = \emptyset$
   then
      foreach $t' \in \left(^F p\right)^{EN}$ **do**
         if $\left|^F t'\right| = 1$ **then**
            if $\left(^F p^* \equiv^{EN} \left(p^F\right)\right) \wedge \left(p^{*F} \equiv^{EN} \left(^F p\right)\right)$ **then**
               include $^F t$ in all $sync\_p \in Sync\_p$;
            end
         end
         else
            foreach $p^* = {}^F t'$ **do**
               if $\left(^F p^* \equiv^{EN} \left(p^F\right)\right) \wedge \left(p^{*F} \equiv^{EN} \left(^F p\right)\right)$ **then**
                  foreach $sync\_p \in Sync\_p$ **do**
                     add $p^* \cup sync\_p$ to New_Sync_p;
                  end
                  $Sync\_p := New\_Sync\_p, New\_Sync\_p = \emptyset$;
               end
            end
         end
      end
      foreach $P^* \in Sync\_p$ **do**
         if $p^F \subseteq^{EN} \left(^F\left(P^*\right)\right) \wedge {}^F p \subseteq^{EN} \left(\left(P^*\right)^F\right)$ **then**
            $event\_sync := event\_sync \cup (p_1^*, \ldots, p_k^*, p)$;
         end
      end
      $Sync\_p := \emptyset$;
   end
   reset all variables;
end

Algorithm 1: Improved algorithm for determination of event-synchronized places.

The Definition 2.4.1 of event synchronization and the related Algorithm 1 enhance the property as defined in the article [MH09]. It includes structures with a set of places synchronizing a place. An example structure is given in Figure 2.8.

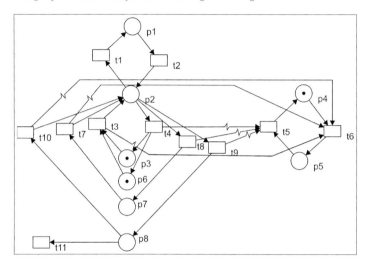

Figure 2.8: Example net for state invariants with synchronizing sets.

The example contains two event synchronizations. First the place $p5$ is synchronized with $p3, p7$ and $m(p5) + m(p3) + m(p7) = 1$ holds. The second state invariant is $m(p5) + m(p6) + m(p7) = 1$. The place $p8$ is not contained in a state invariant, because it has a post-transition ($t11$), which is not triggering a transition around $p5$.

Synchronized places seem to be behaviorally redundant to model parts they are synchronized with, but the systematic and modular modeling as described in Section 4 leads to models containing such structures. The structural property of synchrony can be used to speed up algorithms for analysis and to improve applications of partial markings as the synthesis described in the Chapters 6 and 7. The determination of event synchronization wide the range of state invariants defined without reference on reachability.

Based on the described structures, the property of *linked transitions* can be derived. Transitions are called to be linked if they can fire only together, depending to the initial marking. Such linked transitions can for example be used to expand t- invariants of the underlying Petri nets [SR02]. That is not elaborated further because it does not have affect in the approaches discussed in the following.

2.4.4 Generalized state invariants

Based on the same structural correlation as for event synchronization, another kind of invariant can be defined. Thereby, the term invariant is widened to cover logical predicates on the marking of places holding true for all reachable states.

The conditions for places to be synchronized are very strict. The following described structural predicates generalize them. They can be derived also if some of the conditions of Definition 2.4.1 are not satisfied. If one of the event sinks in the example shown in Figure 2.7 is a sink of a condition signal for example, no synchronized places exist. Such modified example structure is shown in Figure 2.9.

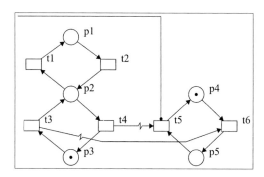

Figure 2.9: Example net containing a generalized state invariant.

However, there can be found predicates on the marking, which hold for all reachable states and can therefore be called to be invariant. Such predicates are called *generalized state invariants* and are determined based on the net structure too. In the example pictured in Figure 2.9 the predicates $\overline{p3} \rightarrow p5$, $p2 \rightarrow \overline{p4}$ and $p1 \rightarrow \overline{p4}$ hold true for every reachable state with the given initial marking. They all are related to the same correlations as the event synchronization but do not satisfy the condition $^{CN}p \cup p^{CN} = \emptyset$.

The example shows just a sample of invariant relations within a model describable by such predicates. It seems to by unfeasible to describe, determine and use all of them. But depending on selective analysis aims, or as side output of already used algorithms, they can improve results. Thus, the example relations in the model in Figure 2.9 can be obtained as side output of the determination of state invariants.

The construction rule of the typ of predicates introduced above can be derived by decomposition of the synchronization of two places. If for two places p and p^* the synchronization $m(p) + m(p^*) = 1$ holds, then it also holds $p \to \overline{p^*}$ and $\overline{p} \to p^*$. A condition signal at an event sink in the post- or pre- region of a synchronized place locks one of those two predicates. The predicate $p \to \overline{p^*}$ depends on the post-transition(s) of the synchronized place p, because it is conditioned by the clearance of the token. The second predicate depends on the pre-transitions, respectively. The predicate is a generalized state invariant if the depending transitions are not conditioned, i.e. they are influenced by a condition signal and all other conditions of Definition 2.4.1 are satisfied. Therefore, the predicate $m(p3) = 0 \to m(p5) = 1$ holds true at the example in Figure 2.9. Under additional consideration of the p- invariants, the predicate $p1 \vee p2 \to \overline{p4}$ is deduced as done with event synchronization. Obviously, there can be found more predicates being generalized state invariants under use of the p- invariants. But they are not extending the knowledge of the behavior. The set of reachable states of the example is completely covered by the p-invariant and the predicate $\overline{p3} \to p5$ under consideration of every state of the condition signal. That only holds if $t5$ is not dead by the state of the condition signal.

Generalized, state invariants can be used for non-reachability checks just like state invariants. But more important, it can be useful in methods based on partial markings to extend the partial markings. The discussed type of invariant predicates can be determined within the algorithm for determination of event-synchronized places.

3 Distributed Safety Control Functions

This chapter characterizes the type of controllers discussed in this work. It introduces their structure and gives a classification within the field of manufacturing control. For this purpose, some major terms are qualified at first. Safety is one of the fundamental terms defining the scope of this work.

Safety is freedom from unacceptable risk of physical injury or of damage to the health of people, either directly, or indirectly as a result of damage to property or to the environment [IEC61508-0].

More specific, this work treats two aspects of *functional safety*. Functional safety is defined as part of the overall safety that depends on a system or equipment operating correctly in response to its inputs [IEC61508-0].

One aspect of functional safety in context of electronic systems is systematic fault prevention. That comprehends all aspects of control as for example controller hardware and software construction. This work deals exclusively with the element of control program design. The increasing size and complexity of (safety) control programs induce the risk associated with such fault to become even more relevant. The automatic *formal synthesis* of controller is such method developed to prevent against faulty influences of human work. Formal synthesis is a method generating controller models proven to be correct mathematically. The base for the synthesis is a formal behavior model of the plant to be controlled and a formal specification of forbidden and/or desired behavior of a controlled plant. The latter is the major difference to many works treating automatic code generation in computer science. The specification does not characterize the controller program but the desired or forbidden behavior of the controlled system. Specifications of forbidden states are used to synthesize safety control function in this work. The synthesized controller model describes the correct controller reaction to achieve the formally-specified behavior of the plant model. This safety specification is the second aspect of safety in this work. It refers to safe operation, while the first one refers to safe design of control functions. The safe design is a prerequisite to safe operation.

Controller synthesis aims to avoid the impact of error proneness of the human work during the program design. This work regards the challenge of inhibiting forbidden behavior and does not deal with process control synthesis. Any plant behavior causing risk to humans, the environment or the equipment is forbidden. The functionality of safety controllers as addressed in this work is beyond the use of safety functions in combination with safety equipment as for example two-hand control devices, guard interlocking and emergency stopping systems. The safety control functions described in this work are supposed to realize any locking or active switching of plant actuators depending on the controller input state. The safety control functions are located as a kind of filter between the process control layer and the plant layer as shown in Figure 3.1.

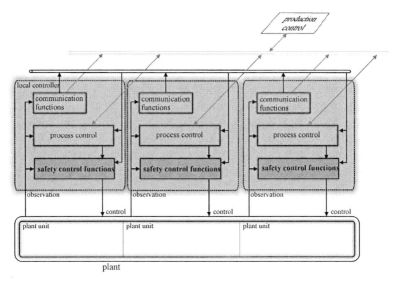

Figure 3.1: Adressed distributed control structure.

This structure regards safety control functions as part of the normal control system. The same systematic can also be used to synthesize and structure the control logic for Safety Instrumentation Systems (abbr. SIS). SIS are specific systems assuring a safe process state even if unacceptable or dangerous process conditions are detected. The complete control loops of a SIS have to be independent from the normal control loops and have to satisfy stronger requirements specified for example in the Standard [IEC61511] for the process industry and [IEC62061] for the manufacturing industry. Those specific requirements and SIS in general are not further addressed in this work.

The result of controller synthesis is a formal controller model. Based on that correct model, the executable controller code can be generated. The controller model consists of predicate functions defining the value of a controller output depending on the input state of the controller. Based on the predicate functions, the executable code can be generated in different programming languages of the IEC 61131-3 [IEC61131-3] or IEC 61499 [IEC61499]. In Chapter 8, a short overview on controller code generation is given and the transformation of synthesized control functions to Basic Function Blocks (BFB) according IEC 61499 using structured text algorithms is discussed. The BFB are chosen because this work focuses on synthesis of distributed controller functions and IEC 61499 especially supports the implementation of distributed controllers. The generated Function Blocks fit between the process controller outputs and the plant inputs as shown in Figure 3.1.

3.1 Distributed control structure

There are different control structures discussed within the field of distributed control synthesis. They differ in the degree of interaction and how interaction is organized. No direct interaction on the controller level exists in distributed controllers without communication as described in [YL02, HVL94]. Such design can be used to design a number of independent controllers based on a global specification. This design also suits for control problems without the ability of communication. Obviously, the distributed controller design with independent observation and control is not as powerful as one homogeneous controller with system-wide observation and controllability. Properties of such distributed controllers and existence criteria are given in [JCK01]. Results of this structure can be applied where no communication between controllers is possible or it is avoided to improve the flexible in the use of production modules.

Cooperation without direct interaction of distributed controllers is described by Yoo and Lafortune [YL02]. They introduce an interconnection structure for the distributed controller outputs. In this structure, output events can be conjunctively and disjunctively combined to define the input for the global plant model. The property of *co-observability* is defined for such distributed controllers. This is enhanced to *conditional co-observability* in [YL04]. None of the mentioned approaches for synthesis of distributed controllers without communication is maximally permissive according to the complete system. A modified approach with interconnection structure is given in [RL03b]. It results in maximally permissive local controllers. The complete system of controllers again is not maximally permissive. An adoption of the results of modular synthesis to decentralized supervisory control of this framework is introduced in [KMP06].

A top-down approach to synthesize local supervisors in a distributed architecture is presented by Cai and Wonham in [CW10]. The local supervisors control specific parts of the plant, but the observation is not restricted to this plant part in contrast to the distributed architecture defined in [MH06b] and used in this thesis. The authors derive the non-interactive local controllers by decomposition of a monolithic supervisor, while its properties are kept. Such a decomposition was introduced for the control problem of safety controllers in [MH06b] and is also used for the synthesis of forcing/locking safety control function in Chapter 7.

A second group of designs are distributed controllers with communication [vS98, BL98, RvS05, IA06]. Controllers interchanging their local observation and controllers interacting via computed variables can be differentiated in this group of designs. Distributed controllers with information structure are defined by van Schuppen in [vS98]. The approach of Lafortune and Barret [BL98] also generates an information structure together with the controllers. Thereby, the controllers also consider which additional prospective estimated information can be used for future decisions in addition to the approach in [vS98]. The generation of a minimal set of communicated events is addressed in [RvS05].

A synthesis approach for distributed controllers exchanging local observation is introduced by Tripakis [Tri04b]. The defined property of *joint-observability* describes the necessary observations to make the decision according to a given specification.

The control theory has been extended by hierarchical control structure within the last years [LBLW01, RW05, LD07, HCdQ$^+$10]. Thereby, model and specification are divided into a number of abstraction levels. Virtual controllers are designed for every layer of abstraction, while the synthesis of each lower layer gains from the behavior restriction given by the layer above. The maximal permissiveness is difficult to prove for the given approaches. A combination of hierarchical and modular structures in a multi-level architecture of supervisors is described in [SB11]. It uses a modular plant model in terms of automata and hierarchically decomposed specifications. The use of the mutual controllability criterion for the lower level allows the synthesis of maximally permissive controllers without composition of the plant modules. The mutual controllability criterion was described in [KvSGM05].

The controllability is restricted by the use of distributed controllers without collaboration. The approaches introduced in this work focus on controllers with communication. More precisely, distributed controllers with communication of state information are synthesized as shown in the controller structure given in Figure 3.1. The controllers exchange state information via communication variables. That means the local controllers evaluate their local observation and send information needed by another controller for decision making. The information structure is strongly linked to the control problem and therefore generated together with the controller. Each approach uses its own method for communication structure design. The modular approach synthesizes the communication structure together with the control functions. The global specification is distributed and the algorithm then proceeds in modular manner. That is shown schematically in Figure 3.2.

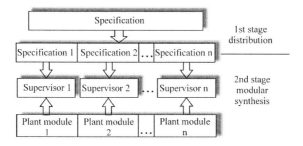

Figure 3.2: Scheme of the modular synthesis.

The monolithic approach instead is performed based on a composed model and a global specification. Figure 3.3 shows the two main stages of the distributed control function synthesis. A preliminary monolithic controller model is distributed in the second stage.

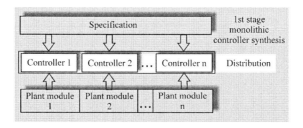

Figure 3.3: Scheme of the monolithic synthesis.

The model type used to define the plant behavior is based on Petri net elements as described in Chapter 2. Distributed controller synthesis based on Petri net based models is
developed in parallel to the already presented approaches based on finite state machines or
language representation. Approaches for *supervision based on place invariants (SBPI)* are
introduced in [YMAL96, MA00]. Based on those works, Iordache and Antsaklis introduced
an algorithm for deduction of distributed controllers in [IA03a]. Both authors describe an
alternative for direct synthesis of the distributed controller in [IA03b]. The synthesis of
a preliminary homogeneous controller is not necessary in this approach. Similarly, both
ways are addressed in the following chapters. Guia and Xie introduce in [GX05] a synthesis
approach based on Petri net unfolding. That avoids the calculation of the whole reachable state space and the resulting controller is non-blocking. A survey on early works on
controller synthesis with Petri nets can be found in [KHG97]. A more recent overview is
given in [IA06]. Different synthesis approaches for homogeneous, modular and distributed
controllers are described in this book.

A direct way of distributed controller synthesis is also introduced by Guan in [Gua00].
This is based on behavior models using so called *condition systems*. That model type is
similar to $sNCES$ but comprehends condition signals as only signal type. The synthesized
controller model structure is divided into blocks for observation, blocks for safety decisions
and task blocks. The local controllers also interact via communication variables.

This work combines the synthesis methods for controllers on the shop-floor level [HLR96,
HTL97, HL98] with the idea to synthesize communication structures for interacting controllers together with the distributed control functions. The synthesized local controllers
consist of local desicion functions (DF) defining the value of the outputs to the plant and
communication functions defining the value of communication variables sent to other local
controllers.

4 Formal Plant Modeling

Any result of a model-based controller design methodology can only be as good as the model of the plant to be controlled. It is therefore a fundamental issue to have a modeling methodology that ensures a clear and concise way of designing formal models for given plants. The following sections thereby focus on plant models for manufacturing systems only. Descriptions and rules for formal modeling of systems, here in particular the behavior of production systems, is discussed less extensive in literature than analyzing methods. Some modeling techniques for discrete event systems for state machine-based models can be found in the books [SC01, Fis07], for algebraic models in [Fok07] and for Petri net models and derivatives in the books [Jen97, RR98, GV03] for example.

Only the plant modeling with $_SNCES$ is addressed in the following because the major focus of this work is on controller synthesis. Previous works on the modeling of $NCES$ models for verification are done by Juhás et al. [JN04] and Vyatkin et al. [VHK$^+$06]. Formal models can also be generated from more abstract and informal models as SysML [OMGSysML1.1], or simulation models for virtual prototyping [PGH10]. Therefore, predefined formal model elements are assigned to abstract model elements and composed as given by the abstract representation. A method for automatic model generation based on simulation models in the X3D data format ([ISO/IEC199775]) is introduced by Karras [Kar09].

The following modeling guidelines are an extension of the earlier works [MH07] and include experiences in model analysis with the later discussed algorithms. At first, the modeled example plant is presented before discussing the formal modeling.

4.1 System example

The system example is part of the modular production system of FESTO. The system consists of four stations for distribution, testing, processing and handling of the work pieces.

distribution   testing   processing   handling
station   station   station   station

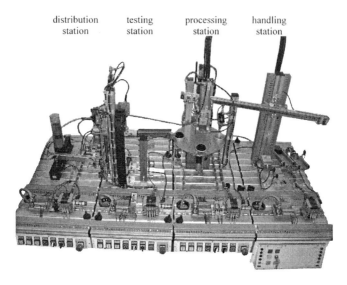

Figure 4.1: Modular production system overview.

Figure 4.1 shows the alignment of the
stations.   More information about the
whole system can be found on the website
[testbed].

The *testing station* is used as system example in this work. The station and its
plant elements are schematically shown in
Figure 4.2. The possible operation is shortly
described as follows.

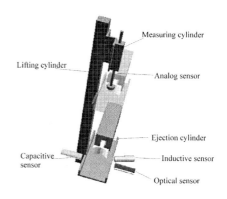

Figure 4.2: Testing station.

The work pieces are provided at the distribution station and individually delivered
to the successive testing station. The testing station is equipped with different sensors
for material properties and the height of the
work piece. A cylinder transports the work
pieces between a lower and an upper level. At the lower level an optical and an inductive

sensor distinguish between red, black, or metal coated work pieces. At the upper level the height is measured by a linear potentiometer sensor with limit switch. Based on that information, work pieces are divided into good pieces for further processing and those that are scrap. The good pieces are rejected on a slide to the handling station at the upper level, while the scrap ones are rejected at the lower level. The presence of a work piece on the lifter is detected by a capacitive sensor equipped at the lifting platform. The cylinders are equipped with end limit position sensors. One sensor is mounted at both end limit positions of the lifting cylinder. Only the extracted position is detected at the ejection cylinder, while the retracted position of the measuring cylinder is observed by a sensor. These two cylinders are spring return cylinders, while the lifting cylinder is not.

Parts of the testing station model are used for demonstration of the algorithms and the whole station is used as benchmark system.

4.2 Well-structured models

Manufacturing systems are completely created by human beings and the complexity mainly is a question of the size of the system. This fact is extremely helpful in modeling since the basic building blocks of the physical manufacturing equipment are always the same. They appear in various combinations but show in general identical or very similar behavior.

The only exception are the goods being processed by the manufacturing equipment. They show individual behavior in each particular case. Obviously, such properties must be included in the model since the goal of the whole manufacturing process is to perform stepwise changes of these properties and a major part of the (process) specifications are expressed in terms of properties of manufactured goods.

It is therefore essential to clearly distinguish partial models describing work pieces, the physical manufacturing equipment modules, sensors, and actuators. The resulting general structure is shown using the example of the lifting unit in Figure 4.3. Models with that structure are called well-structured. The long-term experience shows that coming up with well-structured models without means for modular design is impossible.

It is only natural to build up the plant model as a collection of its basic elements and the interactions between them. Therefore, a set of basic modules for sensors, actuators (valves, electric relays), cylinders, drives and others is used. They are taken from a library for such basic modules. As already mentioned, basic modules are combined to composite modules that represent functional units of the plant. Figure 4.3 shows that composition exemplarily

for the lifting unit. A hierarchical structure is built by combining such composite modules.
The behavior models can also be part of the concept of *automation objects* [Bre05, VHK+06,
Vya06]. The models are then taken from the automation object repository.

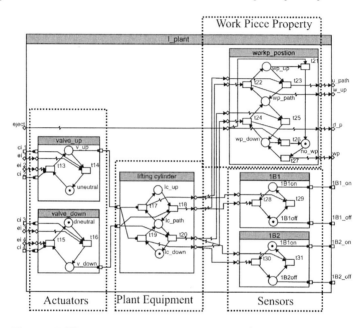

Figure 4.3: Elementary structure of plant models (example *lifting unit*).

The design of well-structured models is illustrated by means of the testing station introduced in Section 4.1. The physical elements to be modeled are three binary property
sensors, the end limit position sensors, and an analog sensor with binary switch, two pusher
cylinders, the pneumatic linear drive (lifter) and its actuating magnetic valves.

The plant model is partitioned into four units. This is one unit for every moving element
and an additional one for the three property sensors. The sensors are not eminently
depending on one of the actuators and are therefore treated in a separate plant unit. The
general design and the hierarchy of the whole plant model are shown in Figure 4.4.

As described before, one composite module is assigned for every plant unit. Basic modules
are built for the physical manufacturing equipment, the actuating relays and proximity

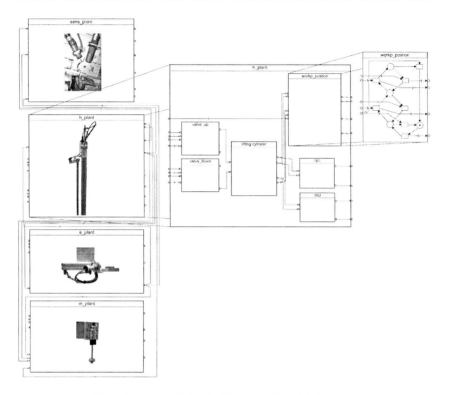

Figure 4.4: Example for the hierarchical model structure.

switches within the module for a unit. Further, a unit model includes modules for the position or/and properties of the work piece that are relevant for that part of the plant. All properties transformed by the plant equipment or detected by sensors of the unit are relevant. Conversely, all signals to sensor inputs have to have their source within the unit, which means that the detected element has to be part of the same unit. The same holds for sensors detecting positions of the equipment as well as for position or properties of the work piece. This regulation supports local analysis of the models as introduced in Section 6. The model of one unit as an excerpt of the whole example is shown in Figure 4.3. It shows the model of the lifter unit consisting of two pneumatic relays for up and down movement of the lifter, the lifter model itself and two modules for proximity switches at the upper and lower end limit position. Furthermore, the position of a work piece at the lifter is modeled by the module *workp_position*. The work piece on the lifter will move if the lifter does. Alternatively, there can be no work piece at the lifter. Hence, the model of the position is similar to the model of the lifter supplemented by a place for *no work piece at the lifter*. The interconnection between the module *lifting cylinder* and the module *workp_position* is given by event arcs since the cylinder enforces a work piece movement immediately if there is a work piece present.

The binary sensor modules *1B1* and *1B2* contain basic two state nets. They are instances of the same basic model type just as the modules for the pneumatic relays. The connections of the proximity switches *1B1* and *1B2* and the lifter are modeled with event signals too. The sensor state is directly depending on the lifter position. In contrast to this, the connections between the relay modules and the cylinder module are condition arcs since the pneumatic coupling has to be regarded as delayed.

The lifting unit module with its six submodules includes just four different module types. Only the work piece property modules have to be modeled individually, while the position model strongly depends on the moving plant equipment models. Such properties are different material, color or geometric shape of a work piece. The states of sensors usually depend on such work piece properties. For example, the behavior of an optical sensor is influenced by the color or material of the work piece. Hence, that interrelation must also exist in the model.

In this work, the labeled signal arcs outside the individually displayed unit modules (as for example Figure 6.5) are part of the interconnection to the other unit modules of the plant (see Figure 4.3). The complete model of the uncontrolled plant is shown in Figure A.3. The uncontrolled plant has 31680 reachable states. The model is built up with the *(T)NCES Editor* developed at the automation technology lab of the Martin-Luther-University [example].

By means of analyzing the modular steps in the model, one can clearly identify the dependencies of the behavior of the modules. This can be useful for model validation, controller synthesis and also controller debugging.

4.3 Controllable and observable plant elements

The behavior of plant elements in real systems is not fully controllable by controllers and accordingly, this also holds for the models. The controllable elements of a plant model are specified by open signal inputs and their interconnections with transitions. Signal sink transitions of such interconnections are defined to be controllable. All controllable transitions can be influenced by condition and event signals. That results from a simplification of the input/output connections between controller and plant and the reverse connections. The elements covered by the simplification are shown schematically in Figure 4.5.

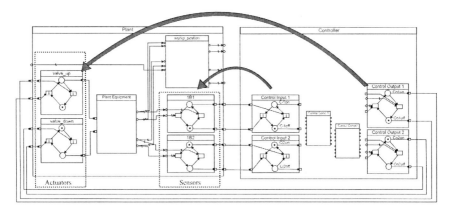

Figure 4.5: Aggregation of control inputs/outputs with plant outputs/inputs.

Controller outputs and plant inputs are merged just like plant outputs and controller inputs. For the synthesis, only a model of the uncontrolled plant behavior is the basis, in contrast to the closed-loop modeling for simulation or verification of the system behavior (see e.g. [PMG+10]). Controller input and output information wouldn't be contained in the model without that merge. It is supposed that control functions exist, which are locking the change of controller output values as well as those functions forcing the change of an output state, as described in Section 3. Therefore, condition and event signals assigned by control functions are modeled influencing the models of the plant actuators. That allows the synthesis of both kinds of control functions, forcing and locking ones.

The modeled plant actuators combine actuator and controller output. The controller output values are defined to be observable by the controller. Hence, the controllable transition's pre- and postplaces are defined to be observable under two conditions. First, the places have only controllable transitions as pre- and post-transitions. Second, the controller's initialization ensures a plant state complying with the initial state of the plant model. That condition is generally to be satisfied for transferability of the model result to

the real world. The state of the controlled plant parts are mapped to the controller output
state if both conditions are satisfied. The set of *controllable transitions* is defined as

$$T^c := \{t \in T : \exists(c^{in}, t) \in CN^{in} \wedge c^{in} \notin CK \vee \exists(e^{in}, t) \in EN^{in} \wedge e^{in} \notin EK\}. \quad (4.1)$$

The set of *controlled places* is therefore defined as

$$P^c := \{p \in P : \{p^F \cup^F p\} \in T^c\}.$$

These controlled places are defined to be observable because of the described merge with
controller outputs. The sensors of a plant are naturally the observable part. The state of
sensors is forced by event signals from the observed plant equipment. Additional conditions (condition interconnections) could be modeled for example if sensor values depend
on a work piece property. The observable plant parts are defined by signal connections
to open signal outputs in the model. Open signal outputs are signal outputs without interconnection to other parts of the plant model. Following, such open condition outputs
and associated interconnections are modeled for the places of sensor models. The *set of
observable places* is defined as

$$P^o := \{p \in P : \exists(p, c^{out}) \in CN^{out} \wedge c^{out} \notin CK\}.$$

This leads to the following short specifications for the model design of the uncontrolled
plant behavior for the controller synthesis:

- Plant parts (like cylinders, conveyors or the like) and work piece properties
  are uncontrollable and unobservable.

- Sensors
  constitute the observable parts of the plant model. They are therefore modeled with
  open signal outputs.

- Actuators (pneumatic or electrical relays and so on)
  are the controllable part of the plant model. The transitions of actuator modules are
  connected with open event and condition signal inputs.

The signal inputs and outputs at controllable and observable parts are unconnected because
only the uncontrolled plant behavior is modeled. That design principle can be seen in the
example in Figure 4.3. Some difference to the plant models for closed-loop modeling of
systems are necessary to include all essential information for synthesis (e.g. two kinds of
control functions, specification of controllable and observable elements).

In summarizing this section, one can state the following:

1. The model structure is a 1 to 1- mapping of the plant structure. This is extremely helpful for designing, debugging and explaining the model to others.

2. Model modules can be used over and over again. This drastically reduces the effort spent in designing the complete model. Designing models of real production systems from scratch is infeasible.

3. The model provides the same signal interface to the controllers as the real plant. This includes limited controllability of state transitions in the plant as well as limited observability of the states in the plant. This reflects reality and is useful for any kind of model-based controller design.

5 Behavior Specifications

5.1 System specification

A formal specification of the required or of the forbidden system behavior is needed for formal synthesis of controllers. The set of forbidden states in terms of predicates is defined for the synthesis of safety controllers. The aim of constructing a safety controller is to prevent the system from reaching situations, which impose a risk to humans, the system, or the environment. Such situation can be specified by forbidden states, forbidden state transitions, forbidden sequences of states or forbidden sequences of state transitions and combinations of them. This work addresses specifications in terms of forbidden states mainly. Specifications in terms of state transitions and also their combinations with forbidden states can be transformed to forbidden state transitions [Lüd00]. The use of sequence specifications results in the need of controller models with memory. Ideas for necessary modifications of synthesis algorithms are given in [Han97].

Dangerous or forbidden states of the physical system are formulated by predicates over the states of the model. The states of the model are represented by the marking of places. The state attributes of a forbidden state are defined as follows:

Definition 5.1.1. (State Atoms/ Predicates/ Attributes)
Let $\mathcal{M}$ be an $_sNCEM$, $p \in P$ be a place of $\mathcal{M}$ and m a marking of $\mathcal{M}$. A state atom ZA of $\mathcal{M}$ at m and p is a declaration:

$$ZA = [m(p) = a]; a \in \{0; 1\}.$$

A state predicate ZP of $\mathcal{M}$ on m is a function of state atoms:

$$ZP = ZA_1 \wedge ZA_2 \wedge \cdots \wedge ZA_n.$$

The set of state atoms contained in a state predicate is

$$\mathbb{Z\!A}^{ZP} = \{ZA_1, ZA_2, \cdots, ZA_n\}.$$

A state attribute ZE of $\mathcal{M}$ on m is a function of state predicates:

$$ZE = ZP_1 \vee ZP_2 \vee \cdots \vee ZP_m.$$

$\square$

For every specified forbidden state, a state attribute is defined. The safety specification of a system (model) is given by the state attribute. A state attribute is satisfied if one of the disjunctive combined state predicates is satisfied. The specification of a safety condition is given as negative specification, because it characterizes forbidden situations.

Definition 5.1.2. (Permissible Behavior)
A trajectory $tr \in TR_{\mathcal{M}}$ is permissible, if there exists no marking $m \in tr$, which satisfies the state attribute ZE. A model of a closed-loop system satisfies the specification, if any trajectories is permissible in the controlled behavior.

$\square$

The controlled behavior is represented by the reachability graph of the closed-loop system. State predicates do not represent a specific marking of the model. Instead, they cover a number of states with the same marking of a subset of places, roughly speaking a partial marking of the model. The use of partial markings allows the analysis of relevant model components only. Places constituting a forbidden state are relevant in the following synthesis algorithms. Further, a notation used in the following chapters is defined to ease descriptions.

Definition 5.1.3. (Predicate Covering)
A state predicate ZP_1 covers a state predicate ZP_2, written $ZP_1 cov ZP_2$, if the following holds:

$$ZP_1 cov ZP_2 = true \; if \; \forall ZA \in ZP_1 : ZA \in ZP_2.$$

$\square$

If a state predicate ZP_2 is covered by a predicate ZP_1, then all states satisfying ZP_2 are satisfying ZP_1 too. That leads to the exclusion of covered predicates in some analysis introduced in the following chapters. The condition is extended to a set of predicates $\mathbb{ZP}$ containing at least one predicate ZP, which covers another predicate ZP'.
For that condition $\exists ZP \in \mathbb{ZP} : ZP cov ZP'$ is shortly written $\mathbb{ZP} cov ZP'$.

The example specifications of forbidden states are outlined in the following. They refer to the system model pictured in Figure A.3.

f1	v_up ∧ v_down	It must not be possible to enable the actuating valves for up and for down movement together.
f2	mc_extend ∧ no_wp_u	The measuring cylinder is not allowed to move if there is no work piece at the upper position.
f3	ec_nr ∧ mc_nr	The measuring cylinder and the ejection cylinder are not allowed to move simultaneously.
f4	ec_nr ∧ lc_path	The ejection cylinder must not move while the lifting cylinder is moving.

All the specified forbidden situations would cause damage to the equipment.

Specification predicates or results in terms of control function predicates have to be distributed to modular sections to form distributed controllers. Whether the specification or the synthesized functions are distributed depends on the approach as can be seen in the Figures 3.2 and 3.3. The distribution of specification predicates is introduced in the following.

5.2 Distribution to modular specifications

Three different kinds of specification state predicates can be characterized in distributed systems. There are local state predicates referencing to just one plant part and depending on the behavior of this part only. The second kind are intermodular specification predicates referencing as well on just one part but depending on the behavior of at least one other plant part. Finally, transmodular specification predicates can be characterized as specification predicates referencing to more than one plant part.

Analyzing algorithms can explicitly deal with the different kinds or avoid a necessary differentiation. The last is realized in the approach presented in Chapter 7. The problem of algorithms explicitly regarding the difference is to discriminate between the local and intermodular specifications. The difference does not show up before the execution of the analysis. That problem is addressed in the modular synthesis presented in Chapter 6. The modular synthesis handles state predicates referencing to one plant part only. Therefore, a transformation of transmodular specification predicates to local/ intermodular specification predicates is described in the following. The term local state predicate covers all state predicates referencing to just one plant part.

The global specifications are decomposed into a conjunction of local state predicates. That is similar to the decomposition of supervisors in a conjunction of modular supervisors used and proven in [RW86].

The definition for the distribution of predicate functions was introduced for the distribution of monolithic control functions to distributed functions in [MH06b].

Definition 5.2.1. (Local State Predicates)
Let $\mathcal{M}$ be an $_SNCEM$ and ZP a state predicate with $\mathbb{ZA}^{ZP}$ the set of state atoms in ZP, then local state atoms are defined as:
$ZA_{M_x} \in \mathbb{ZA}^{ZP} : \forall (m(p) = a) \in ZA_{M_x} | p \in P_{M_x} : M_x \in Mod(\mathcal{M})$.
The local state predicates ZP_{M_x} are:
$ZP_{M_x} = ZA_{M_x1} \wedge ZA_{M_x2} \wedge \cdots \wedge ZA_{M_xn}$,
while the following holds: $\forall ZA \in \mathbb{ZA}^{ZP} : \exists ZP_{M_x}$ with $ZA \in \mathbb{ZA}^{ZP}_{M_x}$.
□

Theorem 3. *For any state predicate ZP holds:*

$$ZP = \bigwedge_{M_x \in Mod(\mathcal{M})} ZP_{M_x}.$$

Proof. Theorem 3 follows directly from the construction of local and general state predicates as conjunction of state atoms and the condition $\forall ZA \in \mathbb{ZA}^{ZP} : \exists ZP_{M_x}$ with $ZA \in \mathbb{ZA}^{ZP}_{M_x}$.
□

Local state predicates represent the local component of a state predicate. That relation has to be considered when local state predicates are used in an analysis. Therefore, an additional type of state predicate representation is introduced. It is defined as consisting of local state predicates and communication variables representing the local predicates of other modules.

Definition 5.2.2. (Modular State Predicates)
Let $\mathcal{M}$ be an $_SNCEM$ and ZP a state predicate and $\mathbb{ZA}^{ZP}_{M_i} = \{ZP_{M_i} | i \in \mathbb{N} : M_i \in Mod(\mathcal{M})\}$ the set of local state predicates of ZP and $ZP_{M_k} = max(ZP_{M_i})$ the ultimate predicate of the set, then a modular state predicate $ZP^M_{M_i}$ to a base module $M_{Bi} \in Mod(\mathcal{M})$ is defined as:

- $ZP^M_{M_0} = ZP_{M_0} \wedge com^+_1 | com^-_0 = ZP_{M_0}$,
- $\forall i \in [1, k-1] : ZP^M_{M_i} = ZP_{M_i} \wedge com^+_{i+1} \wedge com^-_{i-1} | \left(com^+_i = ZP_{M_i} \wedge com^+_{i+1}\right) \wedge \left(com^-_i = ZP_{M_i} \wedge com^-_{i-1}\right)$,
- $ZP^M_{M_k} = ZP_{M_k} \wedge com^-_{k-1} | com^+_k = ZP_{M_k}$.
□

Theorem 4. *For all modular state predicates $ZP_{M_x}^M$ holds:*

$$ZP = ZP_{M_x}^M.$$

Proof. Mathematical induction is used to proof Theorem 4.

Basis

$ZP_{M_0}^M = ZP_{M_0} \wedge com_1^+$

because of $com_1^+ = ZP_{M_1} \wedge com_2^+$ follows

$ZP_{M_0}^M = ZP_{M_0} \wedge ZP_{M_1} \wedge com_2^+$ and because of $com_i^+ = ZP_{M_i} \wedge com_{i+1}$ and $ZP_{M_k}^M = ZP_{M_k}$

follows:

$ZP_{M_0}^M = ZP_{M_0} \wedge \ldots \wedge ZP_{M_i} \wedge \ldots \wedge ZP_{M_k} = ZP.$

Inductive step

$\forall i | 0 < i < k$ holds $ZP_{M_i}^M = ZP_{M_i} \wedge com_{i+1}^+ \wedge com_{i-1}^-$ because of $com_i^+ = ZP_{M_i} \wedge com_{i+1}^+$

and $com_i^- = ZP_{M_i} \wedge com_{i-1}^-$ follows

$\forall i | 1 < i < k-1$ holds $ZP_{M_i}^M = ZP_{M_{i-1}} \wedge ZP_{M_i} \wedge ZP_{M_{i+1}} \wedge com_{M_{i+2}}^+ \wedge com_{M_{i-2}}^-$.

Together with $ZP_{M_k}^M = ZP_{M_k} \wedge com_{k-1}^- | com_k^+ = ZP_{M_k}$ and $ZP_{M_0}^M = ZP_{M_0} \wedge com_1^+ | com_0^- = ZP_{M_0}$ holds forall $i \in [0, k]$

$$ZP_{M_i}^M = ZP_{M_0} \wedge \ldots \wedge ZP_{M_i} \wedge \ldots \wedge ZP_{M_k} = ZP \quad \text{Q.E.D.}$$

□

Local state attributes can be combined of local state predicates as described in Definition 5.1.1 for state predicates. An abstract example is given for the distribution of a transmodular specification covering markings in three different modules in the following. The modules can be numbered /indexed independently of their position or relation within the whole model. Only modules related to the predicate are indexed. The predicate is defined by

$$ZP = ZP_{M_0} \wedge ZP_{M_1} \wedge ZP_{M_2}.$$

The resulting modular specification predicates and their relations are given in Figure 5.1.

The different kinds of arrows show the composition of the modular state predicates by local state predicates represented by the communication variables.

The distribution of the specification in terms of state attributes to modular state predicates has to be performed in the first step of the modular synthesis algorithm as described in Chapter 6. The same rules are also used for distribution of result functions of the monolithic synthesis in Chapter 7. The handling of intermodular specifications is described together with the modular synthesis in Section 6.2.

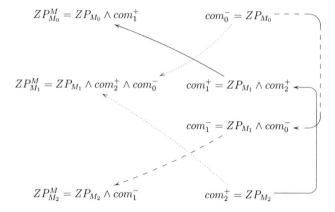

Figure 5.1: Scheme of communication interconnections for distributed specification predicates with k = 2.

6 Modular Synthesis of Locking Safety Controllers

Any synthesis algorithm that copes with systems of real scale must omit the complete enumeration of the state space. The calculation of the reachability graph is obviated because of its consumption of exponential space and consequently also exponential time related to the number of places within the net. The construction of the state space is PSPACE-complete for Petri- nets with the property that no place ever contains more than one token [Val98]. Idem holds for $_SNCES$ if no signal arcs are taken into account. However, signal interconnections relate place markings to each other, especially event interconnections, which can reduce the complexity for some applications.

Therefore, the challenge is to describe an algorithm, which does not work over the complete set of reachable markings. Instead, the following presented algorithms use sets of partial markings generated starting from a given specification state predicate. Thus, only a partition of the reachable state space is analysed starting from the forbidden states. The complexity is further reduced by the use of state predicates covering a number of states.

The idea of the *symbolic backward search* is to analyze the modeled uncontrolled system behavior starting from specified forbidden state predicates. Therefore, the *uncontrollable pre-region* of a predicate defining a forbidden state is identified to prevent the controlled system from reaching forbidden states. The uncontrollable pre-region PR_u is a set of predicates for which exists a sequence of uncontrollable steps leading to a forbidden predicate. It is depicted in Figure 6.1. The backward search algorithm searches for the last possible preventable steps before entering an uncontrollable trajectory to a forbidden state. The uncontrollable pre-region generally is not a subset of the reachable state space because the reachability of states subsumed by the predicates is not reviewed. It is proven to be a superset of the reachable forbidden states in the following.

Uzam and Zhou in [UZ06] used the discrimination into good and bad markings of Petri nets in context of the synthesis of liveness-enforcing supervisors for flexible manufacturing

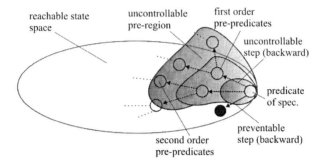

Figure 6.1: Backward search to define the uncontrollable pre-region of a predicate.

systems. Good are all live markings while all markings without a live follower marking are bad in this context. Their approach is improved in [CL11] by Chen and Li to an efficient way to determine a maximally permissive set of control places enforcing liveness. Both approaches determine the first bad place reachable from a good one and are based on representations of the reachability space. This is a kind of the other way around to the backward search determining the first good (not forbidden) marking possibly having a bad one as a follower. Bad are all markings that participate in an uncontrollable trajectory to a forbidden state. However, the approach in [CL11] could be used for the monolithic version of the control problem treated in this work if the forbidden states are modeled as facts as in [Lüd00], and liveness is dedicated as an additional condition. The approach of course would have the price of previous determination of the reachability space. On the other hand the synthesis approaches in this work do not ensure liveness because reachability is not considered.

Approaches similar to the backward search as defined in this work for synthesis of forbidden state controllers starting from a forbidden state predicate were presented in [HLR96, HTL97]. The main difference of the following described algorithms is the backward analysis of steps instead of single transitions. Those earlier approaches did not lead to maximally permissive controller models in general. An approach using the analysis of enabled steps was introduced in [Lüd00] for the synthesis of maximally permissive monolithic locking controllers. The algorithm works on a composed ("flat") $sNCES$ plant model together with a formal specification in terms of predicates defining forbidden states. But, the use of enabled steps annul the advantages of the symbolic backward search because no algorithm exists until now determining the set of enabled steps without exploring the reachability space. The following algorithms claim to combine improved permissive results with efficient execution omitting the state space exploration. Further, they adopt the backward search on the synthesis of distributed controller models. Therefore, first a set of transitions is defined, whose elements constitute steps possibly leading to a forbidden marking, i.e. a marking satisfying a forbidden state predicate.

6.1 Steps under partial markings

The steps used in backward analysis for controller synthesis are different from enabled steps as defined in Definition 2.2.5 and 2.2.7. It cannot be checked if all transitions are marking and condition enabled because of the use of partial markings instead of the marking of the system model and because the steps are analyzed in backward direction. Hence, the steps are defined depending on predicates defining the result of the firing and not the enabling state. Resulting the maximality of steps cannot be verified. A new kind of steps depending on partial markings (the state predicates of Definition 5.1.1) is defined therefore. It is called *possible step* because the step is possibly a subset of the transitions of an enabled step leading to a forbidden marking. A modular definition of possible steps was introduced in [MH08c].

The definition of possible steps is again split into two parts, the *local possible steps* and *possible steps* to support modular analysis. The local possible steps $\xi^p_{\mathcal{M}_x}$ are defined for a module $\mathcal{M}_x$. Further, the possible steps $\xi^p_{ZP_x}$ are defined as the aggregation of local possible steps and depend on a local state predicate $ZP_{\mathcal{M}_x}$. The following definition is enhanced to meet the criterion to improve the permissiveness of the controller models (compared to the definitions in [MH08c]).

Definition 6.1.1. (Local Possible Step)
Let $\mathcal{M}$ be an $_sNCEM$, $\Xi_{\mathcal{M}}$ a set of local steps within $\mathcal{M}$ and $ZP_{\mathcal{M}}$ a local state predicate. A local step $\xi_{\mathcal{M}}$ (following Definition 2.2.5) is a *local possible step* $\xi^p_{\mathcal{M}_x}$ iff holds:
$\nexists t \in \xi^p_{\mathcal{M}_x}$ with:

- $\exists p \in \overline{P} : (p,t) \in F | (m(p) = 1) \in ZP_{\mathcal{M}} \vee \exists p \in \overline{P} : (t,p) \in F | (m(p) = 0) \in ZP_{\mathcal{M}}$
- $\exists p' \in \overline{P} : (p',t) \in CN | (m(p') = 0) \in ZP_{\mathcal{M}} \wedge \nexists (t,p') \in F : t \in \xi^p_{ZP}$.

$\square$

Local possible steps are sets of transitions which are not "marking disabled", i.e. states satisfying $ZP_{\mathcal{M}}$ do not contradict the enabling condition (Definition 2.2.3).

Proposition 6.1.1. *Transitions satisfying the two conditions of Definition 6.1.1 cannot be part of an enabled step ξ with $m(\mathcal{M}) \, [\xi\rangle \, m'(\mathcal{M})$ satisfying $ZP(m(\mathcal{M})) = 0$ and $ZP(m'(\mathcal{M})) = 1$.*

Proof. The theorem is proven condition by condition.

First condition:

Firing a transition t with that condition would remove a token from a place $(m(p) = 1) \in ZP$ or would restore the marking of a place $(m(p) = 0) \in ZP$, according to Definition 2.2.9. From Definition 2.2.3 follows that such place cannot get or lose its marking by another transition within the same step because such transition cannot be enabled. Following, the successor marking cannot satisfy the predicate ZP.

Second condition:

If a place p is unmarked at the post-state and does not have a pre-transition producing a token at the place element of the analyzed step, p is unmarked at the pre-state too (Following Definition 2.2.9). $\rightarrow$ A transition t with $\exists p \in P : (p, t) \in CN | (m(p) = 0) \in ZP$ cannot be condition enabled at pre-state satisfying ZP'. □

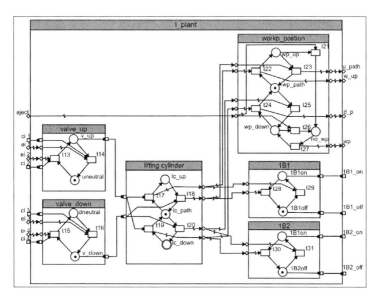

Figure 6.2: Behavior model of the lifting unit.

Local possible steps on their own depend not on a marking but on a local state predicate if it exists for the module. The set of local possible steps is equal to the set of steps if no local state predicate has to be considered.

An *example* for possible steps is given for the model example in Figure 6.2. It models the behavior of the lifting unit. The state atom $ZA = [m(lc\_down) = 1]$ is taken as local state predicate example. The set of local possible steps for the module *lifting cylinder* resulting is $\Xi^p_{liftingcylinder} = \{\{t17\}\,;\{t18\}\,;\{t20\}\}$. The set of steps also includes $\{t19\}$, which is no local possible step because of the first condition in Definition 6.1.1.

Possible steps are defined based on the local possible steps and depending on the modules of the whole modular plant model. The plant models are $_sNCEM$ because they feature signal inputs and outputs for the connection with controller models. The definition can be modified to possible steps in $_sNCES$ by just changing the reference from $\mathcal{M}$ to $\mathcal{M}_S$ in Definition 6.1.2 as defined for steps (Definition 2.2.7, 2.2.8). Possible steps have to satisfy mostly the same conditions as steps. But in difference to Definition 2.2.7, enabled possible steps cannot be defined because no complete marking is considered. Possible steps are defined as follows:

Definition 6.1.2. (Possible Step)
Let $\mathcal{M}$ be an $_sNCEM$ and $ZP_{\mathcal{M}}$ a local state predicate, a step ξ (following Definition 2.2.7), consisting of local possible steps $\xi^p_{\mathcal{M}_x}$, is a *possible step* ξ^p iff the following holds:

Every

$$\xi_{ZP}^p = \bigcup_{\mathcal{M}_x \in Mod(\mathcal{M})} \xi_{\mathcal{M}_x}^p$$

is a possible step if:

- $|\xi_{ZP}^p \cap (\overline{T_{Ei}} \cup \overline{T_{Es}} \cup \overline{T^c})| = 1$,

- for every transition $t \in \xi_{ZP}^p : t \notin (\overline{T_{Ei}} \cup \overline{T_{Es}} \cup \overline{T^c})$ holds:

  - $em(t) = \boxtimes \ \wedge (\exists t' \in \xi_{ZP}^p | t' \rightsquigarrow t)$ or

  - $em(t) = \boxtimes \ \wedge \ (\forall t' \in \overline{T} \ with \ t' \rightsquigarrow t : t' \in \xi_{ZP}^p)$,

- for every transition $t \in \xi_{ZP}^p : t \notin \overline{T^c}$ holds:
  $\nexists p' \in P : p' \rightarrow\!\!\bullet t | (m(p') = 0) \in ZP_\mathcal{M} \wedge \nexists (t, p') \in F : t \in \xi_{ZP}^p$ and

- $\exists \xi_\mathcal{M}^p \in \xi_{ZP}^p : |\xi_\mathcal{M}^p \cap T_{ZP}| \geq 1$, while
  $T_{ZP} = \{t_{p_n}^{pre} \in T : (m(p_n) = 1) \in ZP_\mathcal{M} \wedge (t_{p_n}^{pre}, p_n) \in F\} \cup \{t_{p_n}^{post} \in T : m(p_n) = 0) \in ZP_\mathcal{M} \wedge (p_n, t_{p_n}^{post}) \in F\}$.

Ξ_{ZP}^p is the set of all possible steps depending on the same state predicate ZP.

$\square$

Possible steps are defined depending on a predicate $ZP_\mathcal{M}$ instead of a marking $m(\mathcal{M})$ of the module $\mathcal{M}$.

Theorem 5. *The set of possible steps Ξ_{ZP}^p includes all transition sets depending on $ZP_\mathcal{M}$, which can be part of an enabled step ξ from a marking $m(\mathcal{M}) : ZP_\mathcal{M}(m\mathcal{M})) = 0$ to a marking $m'(\mathcal{M}) : ZP_\mathcal{M}(m'(\mathcal{M})) = 1$.*

Proof. The definition of possible steps is equal to the definition of steps (Definition 2.2.7) constricted by the last two conditions.

The proof of the next to last condition being true for all enabled steps ξ is equal to the proof of the second condition of Propositon 6.1.1.

For the last conditions is stated:

- Following Definition 2.2.7
  $\forall p \in P : (t, p) \in F \wedge t \in \xi | m'(p) = 1$
  $\forall p \in P : (p, t) \in F \wedge t \in \xi | m'(p) = 0$

  $\rightarrow \forall p \in P : m'(p) = 1 \wedge m(p) = 0 | \exists t_p^{pre} \in \xi : (t_p^{pre}, p) \in F$ and
  $\forall p \in P : m'(p) = 0 \wedge m(p) = 1 | \exists t_p^{pre} \in \xi : (p, t_p^{pre}) \in F$

  $\rightarrow \forall \xi \in \Xi : m(N) [\xi\rangle m'(N) \wedge ZP_\mathcal{M}(m(N)) = 0 \wedge ZP_\mathcal{M}(m(N)) = 1 | \exists t_{ZP} \in T_{ZP} : t_{ZP} \in \xi$.

Following, all enabled steps from $ZP_\mathcal{M}(m(N)) = false$ to $ZP_\mathcal{M}(m'(N)) = true$ satisfy those two conditions too. The set of enabled steps is a subset of the set of steps by Definition 2.2.7. The set of possible steps is a subset of steps based on the analogies between their definitions, restricted by conditions satisfied by enabled steps ξ.

Hence, there cannot exist any transition set not in Ξ_{ZP}^p but in an enabled step ξ and influencing the marking of a place p with $ZA = [m(p) = a] \in ZP$. $\qquad\square$

For the **example** in Figure 6.2 and considering the state predicate $ZP = lc\_down$, every possible step has to include the transition $t_{ZP} = t20$ because it is the only one being able to produce a token at the place lc_down. The set of possible steps Ξ_{ZP}^p at the $_sNCEM$ l_plant therefore is $\Xi_{ZP}^p = \{\{t20\} ; \{t20, t25\} ; \{t20, t31\} ; \{t20, t25, t31\}\}$. Only the step $\{t20, t25, t31\}$ would be enabled under the example marking given in Figure 6.2.

The determination of possible steps is discussed together with the synthesis algorithm in the next section.

6.2 Modular backward search

The backward search algorithm determines the complete set of uncontrollable trajectories leading to a forbidden state. The main advantage of the approach presented in the following is that the backward search is performed for every module of a modular plant model separately. Earlier approaches using backward search required fully composed plant models. The model-wise analysis requires additional handling of dependencies between the behavior of different modules resulting from signal interconnections. The idea of a modular execution of backward search was first introduced in [MH08a, MH08c].

The algorithm takes advantage of the modular model and generates local pre-predicates to all local possible steps in a possible step to a local forbidden state predicate for the modular synthesis. Therefore, local possible steps are determined, and the pre-predicates enabling those possible steps are generated. The local possible steps of a module M_x are numbered by the index o in the following. The analysis of the steps starts at that module the processed local state predicate refers to. Pre-predicates are determined for local possible steps leading to a marking satisfying the local state predicate. The pre-predicates have to characterize all states enabling the possible step, that means all states satisfying the *enabling rule*. The enabling rule is stated to be:

Proposition 6.2.1. *(Enabling Rule)*
A step ξ can be enabled only if:

- $\forall p \in P$ with $(p,t) \in F | t \in \xi : m'(p) = 1$
- $\forall p \in P$ with $(t,p) \in F | t \in \xi : m'(p) = 0$
- $\forall p \in P$ with $(p,t) \in CN | t \in \xi : m'(p) = 1.$

Proof. Proposition 6.2.1 follows directly from the Definitions 2.2.3, 2.2.7 for the enabling of transitions and enabled steps. □

If the analyzed local step is uncontrollable, the pre-predicates are part of the uncontrollable pre-region of a specified forbidden state and identified as forbidden states too.

The pre-predicates $ZP'_{\mathcal{M}}$ for every local possible step contained in a possible step ξ^p_{ZP} belonging to a predicate $ZP_{\mathcal{M}}$ are calculated due to the following rules:

The *re-determination rule* for $ZP'_{\mathcal{M}}$ is:

- $\forall p \in P : \exists t \in \xi^p_{\mathcal{M}} : (p,t) \in F \Rightarrow ZA' = (m'(p) = 1) \in ZP'_{\mathcal{M}},$
- $\forall p \in P : \exists t \in \xi^p_{\mathcal{M}} : (t,p) \in F \Rightarrow ZA' = (m'(p) = 0) \in ZP'_{\mathcal{M}},$ notation $\overline{ZA'}$,
- $\forall p \in P : \exists t \in \xi^p_{\mathcal{M}} : (p,t) \in CN \Rightarrow ZA' = (m'(p) = 1) \in ZP'_{\mathcal{M}}.$

The statements follow from the marking and condition enabling conditions for transitions (Definition 2.2.3) and the Definition 2.2.7 of enabled steps.

In difference to the monolithic approach, for example in [MH06b], additionally a rule considering condition couplings is needed.

Therefore the *condition enabling rule* is defined for all $ZP'_{\mathcal{M}}$:

$$\forall p \in P_{\mathcal{M}_x} \text{ with } \mathcal{M}_x \in Mod(\mathcal{M}) \wedge \exists t \in \xi^p_{\mathcal{M}_x} : (p \mathbin{-\!\!\bullet} t) \Rightarrow ZA' = (m'(p) = 1) \in ZP'_{\mathcal{M}_x}.$$

That rule follows directly from Definition 2.2.3 too. It has to be ensured that transitions are condition enabled by condition inputs and therefore by the source places of condition interconnections. Condition inputs without connection to a source place are not considered by the rule. The open signal inputs are not considered in the rules, and the transitions are handled as spontaneous firing transitions simulating an arbitrary input state. The open inputs only define controllable transitions.

Further, the following *propagation rule* has to hold:

For a state atom ZA or $\overline{ZA}$ of $ZP_{\mathcal{M}}$ and $\xi_{\mathcal{M}}^p$ a local possible step from $ZP'_{\mathcal{M}}$ to $ZP_{\mathcal{M}}$ the atom ZA or $\overline{ZA}$ is in $ZP'_{\mathcal{M}}$ too iff for p with $ZA = [m(p) = a]$ holds:

$$\nexists t \in T \, with \, (p,t) \in F : t \in \xi_{\mathcal{M}}^p \wedge (m(p) = 0) \in ZP_{\mathcal{M}} \text{ and}$$

$$\nexists t \in T \, with \, (t,p) \in F : t \in \xi_{\mathcal{M}}^p \wedge (m(p) = 1) \in ZP_{\mathcal{M}}.$$

The predicate $ZP_{\mathcal{M}}$ can be empty for local possible steps. This holds for example if the local possible step is within a possible step depending on a local state predicate of another module. The propagation rule follows directly from Definition 2.2.9. This definition defines that the marking of places without transitions $t \in \xi$ in post- and pre- region remains unchanged by ξ.

The predicate $ZP'_{\mathcal{M}}$ has to be free of contradictions. Thus, $ZP'_{\mathcal{M}}$ has to satisfy the following *consistence property*:

There must not exist two state atoms ZA_1 and ZA_2 within $ZP_{\mathcal{M}}$ which contradict each other, i.e. it must hold:

$$ZA_1 = (m(p) = a) \wedge ZA_2 = (m(p) \neq a), \text{ while } a \in \{0; 1\}.$$

Further, every local pre-predicate must not contradict a state invariant.

Pre-predicates with contradictions cannot enable possible steps, and the possible steps following are never enabled and not taken into account further on. The same holds if a state predicate contradicts a state invariant by definition.

The local pre-predicates $ZP'_{\mathcal{M}}$ are determined under consideration of the mentioned rules. Thereby, pre-predicates are generated for all local possible steps included in the possible step belonging to ZP.

Definition 6.2.1. (Local Pre-Predicate)

Let $\mathcal{M}$ be an $sNCEM$, $ZP_{\mathcal{M}}$ a local state predicate and ξ_{ZP}^p a possible step over the analysed composite module $\mathcal{M}_C$ with $\mathcal{M} \in Sub(\mathcal{M}_C)$. For every local possible step $\xi_{\mathcal{M}_x}^p \subseteq \xi_{ZP}^p$ the set of pre-predicates $PZP_{\mathcal{M}}(ZP_{\mathcal{M}}, \xi_{\mathcal{M},x}^p)$ of $ZP_{\mathcal{M}}$ is the set of state predicates $ZP'_{\mathcal{M}}$, which is calculated by backward analysis of steps $\xi_{\mathcal{M}}^p$ under consideration of the re-determination rule, the condition enabling rule, the propagation rule and the consistence property. $\square$

The pre-predicates as well as the specified local forbidden state predicates are extended by event synchronizations (see Section 2.4.3), p-invariants (see Section 2.4.2) and p-invariants with substituted event synchronizations to improve the backward synthesis results in terms of included observable state atoms. Further, predicates covering another predicate can be detected more likely. The substitution of a place within p-invariants by the negated or complement place synchronized to it can lead to new state invariants not covered by a p-invariant or event synchronization. Examples are given in Table 6.1.

Additional communication variables com_l are generated to consider the dependency between the local pre-predicates of the local possible steps of the same possible step. The l is the index of the variable. A pair of two communication variables com_l^+ and com_l^- is defined for every event interconnection necessary for building the possible step. That means that one pair for every ek satisfying the second condition of the Definition 6.1.2 of possible steps is generated. A communication variable representing communication in direction of the event interconnection ek is symbolized with com_l^+, and the one in reverse direction with com_l^-, as displayed in Figure 6.3. A pair of communication variables belonging to an event chain $t \rightsquigarrow t'$ is symbolized by $com_l \rightleftharpoons (t \rightsquigarrow t')$.

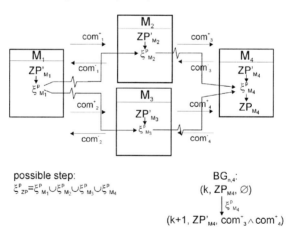

Figure 6.3: Example scheme of a modular backward step and it's results.

The local possible steps are attached with the communication variables to event chains included in those possible steps the local possible step is part of. Therefore, *local communication predicate components* $coc_{\xi^p}^{M_x}$ are defined for each module $\mathcal{M}_x$ a local possible step $\xi_{\mathcal{M}_x}^p \subseteq \xi_{\mathcal{M}_x}^p$ exists within.

A $coc_{\xi^p}^{\mathcal{M}_B}$ is built $coc_{\xi^p}^{\mathcal{M}_B} = coc_{\xi^p}^{\mathcal{M}_B} \wedge com_l^{(+,-)}$

- for all com_l^+ for which holds: $\exists \xi_{\mathcal{M}_B}^p \subseteq \xi_{ZP}^p \wedge \exists t \in \xi_{ZP}^p : t \rightsquigarrow t_{\mathcal{M}_B}^t \wedge t_{\mathcal{M}_B}^t \in \xi_{\mathcal{M}_B}^p \wedge t \notin \xi_{\mathcal{M}_B}^p$,

- for all com_l^- for which holds: $\exists \xi_{\mathcal{M}_B}^p \in \xi_{ZP}^p \wedge \exists t_{\mathcal{M}_C}^t \in \xi_{ZP}^p : t \rightsquigarrow t_{\mathcal{M}_C}^t \wedge t \in \xi_{\mathcal{M}_B}^p$.

The local possible steps $\xi_{\mathcal{M}_x}^p$ and the $coc_{\xi^p}^{\mathcal{M}_x}$ form the local step information $LS(\xi_{\mathcal{M}_x}^p, coc_{\xi^p}^{\mathcal{M}_x})$.

An *example* structure and its resulting step is given in Figure 6.3. The $LS = (\xi_{M_4}^p, coc_{\xi_{ZP}^p}^{M_4})$ is the local step information for M_4 resulting from ξ_{ZP}^p.

The generated local pre-predicates together with the local possible steps can be represented as graphs called *local backward graph*, where $BG_{n,x} = (CP_{\mathcal{M}_x}, BA)$ is the nth graph within the module $\mathcal{M}_x$. The nodes are $CP_{\mathcal{M}_x} = (k, ZP, CO)$ with the index k, the local state predicate $ZP_{\mathcal{M}_x}$ and an associated communication predicate CO. The arcs are $BA = \{ZP_{\mathcal{M}_x}, \xi_{\mathcal{M}_x}^p, ZP_{\mathcal{M}_x}'\}$, while $\xi_{\mathcal{M}_x}^p \subseteq \xi_{ZP}^p$. For every newly assigned pre-predicate it is checked whether there already exists a node with the same predicates or not. If not, a new node and the depending arc are created. Thereby, CO is inherited from $ZP_{\mathcal{M}_x}$ the local possible step is determined for if it exists. A backward step to an already existing node is considered within CO of the existing node in a way described later on.

The communication variables are considered in the local backward graph, more precisely in the communication predicate CO. The predicate consists of all communication variables related to the forbidden state predicate. Every determined predecessor node inherits the CO of the successor node belonging to the analyzed possible step. The new local communication predicate to ξ_{ZP}^p is attached conjunctively $CO := (CO) \wedge coc_{\xi^p}^{\mathcal{M}_x}$. The scheme of defining nodes and arcs of a local backward graph to a possible step ξ^p is shown in Figure 6.3.

If a local possible step $\xi_{\mathcal{M}_x}^p$ is enabled by a local pre-predicate ZP' for which a node (k, ZP_k', CO_k) already exists, then the communication predicate CO of the successor node of the step is connected conjunctively with $coc_{\xi_{\mathcal{M}_x}^p}^{\mathcal{M}_x}$, and the resulting predicate is attached disjunctively to the existing CO' $(CO' := CO' \vee (CO \wedge coc_{\xi_{\mathcal{M}_x}^p}^{\mathcal{M}_x}))$.

Example:

The association of state predicates and communication variables is schematically illustrated in Figure 6.4. The step $\xi_{\mathcal{M},2}^p$ in Figure 6.4 is such a step enabled by an already determined state predicate ZP_3 contained in the existing node $(3, ZP_3, CO_3)$. Thus, the communication predicate $CO_3 = CO_1$ is extended by the predicate $coc_{\xi_{\mathcal{M},2}^p}^{\mathcal{M}}$ resulting from $\xi_{\mathcal{M},2}^p$ to

$CO_2 \wedge coc^{\mathcal{M}}_{\xi^p_{\mathcal{M},2}}$. It has to be mentioned that only nodes with identical state predicates or nodes covered by a predicate with identical CO are melt that way. A node capturing a subset of states of another node has to be treated separately. Otherwise, the control gets unnecessarily restrictive. That difference to the later described monolithic approach results from the fact that only modular predicates are handled. The global predicates can differ in other predicates represented by communication variables and the global predicate could be not covered therefore.

In Figure 6.4, the arcs are directed in the firing direction of steps. The backward search algorithms analyzes the steps in reverse direction.

Generally, the composition of communication variables is deduced from the representation of backward search in a monolithic model. Every conjunction of communication variables composes one predicate over the composed model. Disjunctive coupling represents different "global" predicates with the same local component. That relation naturally follows as conversion from the proven rules for distribution of monolithic predicates presented in [MH06b] and retreated in Section 7.1.2.

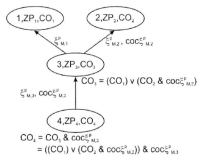

Figure 6.4: Schematic example of the creation and the composition of communication predicates.

If a communication predicate to a local state predicate ZP_k (e.g. ZP_3 in Figure 6.4) is extended because of a new connection within BG, the CO's have to be extended by all nodes contained in the trajectory enabling the state predicate ZP (as can be seen at node 4 in Figure 6.4). This means all CO of nodes $BG_{n,x} = (g, ZP_g, CO_g)$ with $\exists \{BA\} : (g, ZP_g, CO_g) \left[\xi^p_{\mathcal{M}_x,1}, \ldots, \xi^p_{\mathcal{M}_x,n} \right) (k, ZP_k, CO_k)$ are extended.

The control functions of the distributed controllers are generated directly by analyzing the local backward graphs.

The use of graphs including the communication predicates gives the possibility to use the results of already analyzed local step trajectories for all in the same run specified forbidden states. Different forbidden states can depend on the same behavior of the plant or at least on the same local behavior of a plant unit. Therewith, the analysis of different forbidden states can be accelerated by the combination with an already analyzed local state predicate. Every generated local pre-predicate is checked for being already analyzed to prevent repeated treatment. If such an already analyzed predicate occurs, an arc from the analyzed local predicate (ZP) is generated to the local pre-predicate (ZP') at BG, and a modification of the communication predicate of the affected nodes is processed in the already described manner. Two graphs can be melt that way to one if they contain local steps leading to the same pre-predicate (see Figure 6.4).

The termination criterion for the backward search path is the detection of controllable steps to a forbidden state predicate. Hence, it has to be checked for every calculated possible step if the trigger transition is a controlled one ($\exists t \in \xi_{ZP}^p : t \in T^c$). Such steps can be prevented from firing by disabling the condition input of the plant, this means the step is controllable. If a calculated step can be prevented by setting a condition input to zero, then the pre-predicate is stored together with the locking inputs as decision predicate DP. The required condition input ($c_o^{in} = \{c^{in} \in C^{in} \mid \exists t \in \xi_{\mathcal{M},o}^p : (c^{in}, t) \in CN^{in}\}$) for the local step from $ZP_{\mathcal{M}}'$ to $ZP_{\mathcal{M}}$ is identified therefore. This results from the the enabling rule for steps in $_sNCES$. If a marking occurs satisfying $ZP_{\mathcal{M}}'$, then an input status is' would have to enable the step $\xi_{\mathcal{M}}^p$ to make a marking satisfying $ZP_{\mathcal{M}}$ reachable and for is' has to hold:

$$\forall c^{in} \in C^{in} \; : \; \exists t \in \xi_{\mathcal{M}}^p \; : \; (c^{in}, t) \in CN^{in} \wedge c^{in} \notin CK \Rightarrow is' = 1.$$

The controller has to prevent the firing of $\xi_{\mathcal{M}}^p$ to avoid reaching the forbidden state. The control function has to ensure:

If the system is at a state satisfying $ZP_{\mathcal{M}}'$, then $c_i^{in} \in C^{in} \mid \exists t \in \xi_{\mathcal{M}}^p \; : \; (c_i^{in}, t) \in CN^{in} \wedge t \in T^c\}$, $is'(c_i^{in}) = 0$ has to be true for the controllable input.

The node to the pre-predicate $ZP_{\mathcal{M}}'$ and the input to be influenced are stored at a set of decision components $DP = \{(CP_{\mathcal{M}}, c_i^{in})\}$. They are the control components resulting from a controllable step. The modular backward search with the communication variable handling is described by the pseudo code Algorithm 2.

Algorithm 3 checks the controllability of the analyzed possible step and determines the decision predicate for controllable steps. This is used in Algorithm 2.

while $!(LS = \emptyset \text{ and } FSP = \emptyset)$ **do**

  if *the set of LS is not empty* **then**

    forall the $LS(\xi_M^p, coc_{\xi_{ZP}^p}^M)$ **do**

      determine ZP' to ξ_M^p, extend by state invariants and check for conflicts;

      if $\exists BG_{n,x} = (k, ZP_k, CO_k) : ZP = ZP'$ **then**

        bind $coc_{\xi_{ZP}^p}^M$ with $CO \Rightarrow CO_k \vee coc_{\xi_{ZP}^p}^M$;

        bind $coc_{\xi_{ZP}^p}^M$ with the CO of all pre-nodes $BG' = (l, ZP_l, CO_l)$ for which

        holds $\exists w^p | ZP_l [w^p\rangle ZP_k$;

      else

        define new graph $BG_{o+1,x} := (1, ZP', CO := coc_{\xi_{ZP}^p}^M)$;

      end

      check controllability (Algorithm 3);

    end

  else

    forall the $BG_{m,x}(k, ZP, CO) : ZP \in FSP$ **do**

      determine the set Ξ_{ZP}^p of all possible steps ξ_{ZP}^p leading to ZP;

      forall the $\xi_{ZP}^p \in \Xi_{ZP}^p$ **do**

        determine the communication variables $coc_{\xi_{ZP}^p}^M$ to ξ_{ZP}^p;

        pass all $LS_j(\xi_{M_j}^p, coc_{\xi_{ZP}^p}^{M_j})$ to the modules $M_j \to LS_{M_j} := LS_{M_j} \cup LS_j$;

        determine ZP' to ξ_M^p, extend by state invariants and check for conflicts;

        if $\exists BG_{k,x} = (l, ZP_l, CO_l) : ZP_l = ZP'$ **and**

        $(\exists DP : CP_{k,x} \in DP \wedge |\xi_M^p \cap T^c| = 1 \vee \nexists DP : CP_{k,x} \in DP \wedge |\xi_M^p \cap T^c| = 0)$

        then

          bind $CO_k \wedge coc_{\xi_{ZP}^p}^M$ with $CO_l \to CO_l := CO_l \vee CO_k \wedge coc_{\xi_{ZP}^p}^M$;

          bind $CO_k \wedge coc_{\xi_{ZP}^p}^M$ with all pre-nodes to (l, ZP_l, CO_l);

          merge $BG_{m,x}$ and $BG_{k,x}$ by arc $((k, ZP_k, CO_k), \xi_M^p, (l, ZP_l, CO_l))$;

        else

          add $(k + 1, ZP, CO := CO \wedge coc_{\xi_{ZP}^p}^M)$ to $BG_{m,x}$;

        end

        check controllability (Algorithm 3);

      end

      $FSP := FSP/ZP$;

    end

  end

end

delete all nodes $CP = (k, ZP, CP)$ not sink of an arc (ZP', ξ_M^p, ZP) and not in a DP;

check controllability;

    Algorithm 2: Pseudo code algorithm for the modular backward search.

if $|\xi_M^p \cap T^c| = 0$ **then**
$\quad \big|\quad FSP := FSP \cup ZP'$;
else
$\qquad$ check observability (execute Algorithm 4);
$\qquad$ construct a decision predicate DP of the last node determined by Algorithm 4 or
$\qquad$ of the pre-node of ξ_M^p if no extension exist;
end

Algorithm 3: Sub-algorithm of Algorithm 2 checking the controllability and determining the control functions.

The result of Algorithm 2 is a set of local backward graphs for every module with behavior related to the specified forbidden states. The whole synthesis process including the final extraction of the control functions is presented in the next section. Following, the synthesis with modular backward search is illustrated by an example.

6.3 Modular synthesis of distributed locking control functions

In contrast to the monolithic synthesis (introduced in [MH06b]), the backward search is performed module by module under consideration of the signal interconnection between them. The result is a distributed control structure with one controller for every unit module. Therefore, the control structure is predefined by the modular structure of the plant model. The synthesis procedure runs through the following steps:

1. composition of the hierarchical modular plant model to a modular structure of basic modules for the units,

2. derivation of local specifications from a global specification,

3. determination of structural properties (event synchronization, p-invariants),

4. modular backward search and transformation to observable predicates,

5. derivation of the control functions from the backward search results.

These steps are described in more detail in the following.

An $_sNCES$ plant model can consist of multiple hierarchy levels. It is vertically composed to a composite plant module containing basic modules only. Composition is described in

Section 2.3. Every basic plant module then represents the observed and controlled plant part for exact one controller.

Next, the forbidden state predicates have to be *transformed to local state predicates*. The methodology is similar to the distribution of the control predicates presented in [MH06b]. If a specification predicate includes state information of more than one plant module, all state atoms related to the same module are cut out from the state predicate and form a local state predicate as defined in Definition 5.2.2. Every local predicate has to be connected to at least one other local predicate from the same forbidden state predicate by a communication variable. The number of communication variables additionally has to be minimal. If a local state predicate is assigned to more than one communication variable, then the conjunction of the variables builds the CO. The assignment of communication variables can be done in any order under these conditions. The building of the CO is similar to the representation of event interconnection within a possible step described in Section 7.1.2. The local forbidden state predicates are extended by p-invariants, event synchronizations and p-invariants substituted by event synchronizations. An example for building local specifications is presented later in this section.

The *modular backward search* is started as described in Section 7.1.2 using the local state predicates and the modular plant model. The backward search will terminate when for every forbidden state predicate only controllable backward steps are calculated or no more pre-predicates are determined (i.e. if only contradicting pre-predicates exist). The backward search results in state predicates including observable and non-observable plant places. Implemented control functions must include only observable plant elements because only their state can be evaluated by the controller through the plant output signals (observable sensor states) and the own output state (controllable actuator states). The controllable and observable plant elements are identified following the rules introduced in Section 4.3. The state predicates of the nodes of the local backward graphs are transformed to predicates containing state atoms of observable places only, in extension of the approach presented in [MH08c]. A method for the transformation was introduced in [MH09].

The main idea is to take all non-observable state information as always satisfied. That means all state atoms of non-observable places are substituted by Boolean true in the control functions. Obviously, the result can be reduced to functions ZP^o including only observable state information. All controllable model parts are also interpreted as observable (see Section 4.2). Every state predicate ZP of the nodes of every local backward graph is transformed into an *observable state predicate* ZP^o with:

$$ZP^o := \bigwedge ZA : ZA \in ZP \wedge ZA = [m(p) = a] \wedge p \in \{P^o \cup P^c\}.$$

Local backward graphs can include nodes with state predicates containing no observable
state atom. Such non-observable state predicates cannot be evaluated by the controller
and have to be assumed to be true within the safety controllers. They are replaced by the
value logical true in the graph nodes and decision functions.

The synthesis results under consideration of event synchronization consist of a more concrete partial marking than without, and therefore also more concrete observable partial
markings. Only event sources to the analyzed transition t are reasonable for firing of t
and included in possible steps. Therefore, only the sources are considered during the construction of possible steps. The event synchronization includes transitions influenced by
the transitions of a possible step (i.e. some of their event sinks) and their effect on the
marking into the extended pre-predicates. A comparison of the backward search results
with and without extension using event synchronizations was presented in [MH09].

Pre-predicates of controllable steps are always observable because the controlled places are
defined to be observable too. Control functions including only actuator information are
not adequate solutions of the control problem in every case. They are restrictive because
no "real" plant observation is included. An extension of Algorithm 3 avoids such control
functions if possible. It simply proceeds further backward search onto an alternative pre-
predicate. If there is a pre-predicate to ZP' containing also "real" plant observation and all
therefore determined predicates are not satisfied by the initial marking, then the alternative
is used as control component. The algorithm extension is shown in Algorithm 4.

---

if $\{ \bigcup_{p \in ZA: ZA \in ZP'} p\} \cap P^o = \emptyset$ **then**

    search for an alternative pre-predicate ZP'' (execute Algorithm 2 with

    $FZP := ZP'$);

    if $\{ \bigcup_{p \in ZA: ZA \in ZP''} p\} \cap P^o \geq 1$ *and* $\forall ZP^* : ZP'[\xi_1^p \ldots \xi_{n-i}^p) ZP^*[\xi_{n-i+1}^p \ldots \xi_n^p) ZP''$ [1]

    holds $ZP^*(m_0) = false$ **then**

        extend $BG_{m,x}$ by the determined backward graph to ZP';

    end

end

    Algorithm 4: Extension for Algorithm 2 for incomplete state observation.

---

[1] $[\xi_{n-i+1}^p \ldots \xi_n^p)$ denotes a sequence of steps.

When the backward search is completed and the nodes of the resulting graphs are transformed to contain only observable state atoms, the *control functions* are generated from the local backward graphs. The distributed controllers consist of local decision functions defining the state of control outputs and communication functions defining the value of the communicated variables. The structure is introduced in Figure 3.1. Only nodes of BG's with a decision predicate DP are of interest for the decision functions. One function is defined associated to every plant input part of a DP. The functions are built locally because every plant part input is controlled by exactly one local controller, i.e. by analyzing the BG's of one module. *Local decision functions* $DF_{c^{in}}^{\mathcal{M}_x}$ are defined for every local input c^{in} as

$$DF_{c^{in}}^{\mathcal{M}_x} = \bigvee_{\forall ZP_{\mathcal{M}_x} \in (CP_{\mathcal{M}}, c^{in}) \wedge (CP_{\mathcal{M}}, c^{in}) \in DP} (ZP_{\mathcal{M}_x} \wedge CO_{ZP_{\mathcal{M}_x}}).$$

The decision function consists of local state predicates and communication variables. The communication variables represent a set of local state observations of other (i.e. not directly observable) plant parts. In [RW86] was shown that global predicates can be decomposed to conjunctively combined local predicates. Therefore, local state predicates are conjunctively combined with communication variables representing "external" state predicates.

Communication functions are generated for every communication variable included in a communication predicate. Before the functions are generated, the communication predicates CO are transformed to their (minimal) Disjunctive Normal Form (DNF) CO^{DNF}. The Quine-McCluskey method (e.g. in [Shi98, Mar03]) or Thelens method (e.g. in [Kar07]) for Conjunctive Normal Form (CNF) as the source can be used for the transformation to minimal DNF. Every conjunction in the DNF of a communication predicate represents a combination of communication variables defining together a single system state, while the disjunction represents different global states including the same local predicate. The following rule for generation of communication functions results from that context.

The *communication function* CV_l defining the value of com_l is a disjunctive composition of all state predicates associated with a CO including the variable com_l. The communication predicates are built from the local state predicates $ZP_{\mathcal{M}_x}$ combined conjunctively with a communication term. That term consists of the disjunctive combination of all conjunctive terms of CO including com_l, while com_l has to be removed from these terms.

$$CV_l = com_l^s := \bigvee_{\forall BG_{\mathcal{M}}} \bigvee_{\forall CP \in BG_{\mathcal{M}} : com_l^{\neg s} \in CO \wedge CP \notin DP} \left(ZP_{\mathcal{M}_x} \wedge \bigvee_{\forall coc : com_l^{\neg s} \in coc} coc \backslash com_l^{\neg s} \right),$$

while coc are the conjunctive terms of CO^{DNF} and $s \in \{+, -\}$.

Because the communication variables are pairs representing the same connection in different directions, communication functions defining the complementary variable to all variables included in the CO of the node are generated for every node within BG. Thus, a communication function (e.g. CV_{l-} with $com_l^- := CV_{l-}$) consists of a disjunctive coupling of state predicates of all those nodes its complementary variable (com_l^-) is part of.

The result is a set of local decision functions DF and communication functions CV for every plant part. These functions together form the distributed controllers, as Figure 3.1 shows. It is the same structure of local controllers and communication variables as presented in [MH06b] for a monolithic synthesis approach. But, the functions are got from the backward search result directly in contrast to the monolithic approach. The number of communication variables generated is less or equal compared to the monolithic approach. The communication functions from the modular approach can contain chains of communication variables, i.e. communication functions can include communication variables. This structure can lead to a reduced complexity of the communication between the local controllers. The computational complexity of the synthese is reduced by the use of modular backward search compared with the monolithic synthesis. Every local state predicate is determined just once independent of its global context. That context is handled within the communication predicates only. In contrast, a local state predicate can be contained in different global state predicates, all backward analyzed separately using the monolithic approach. These advantages are pointed out using the following example.

Example:
The modular distributed controller synthesis is exemplified in the following using a cutout of the system example described in Section 4.1 and a specification of two forbidden state predicates $f3 = (ec\_nr \wedge mc\_extend)$ (the ejection cylinder must not move, if the measuring cylinder is extended) and $f4 = (ec\_nr \wedge lc\_path)$ (the ejection cylinder must not move while the lifting cylinder is moving). The specification covers the lifting unit, the ejection unit and the measuring unit. The model of the whole system example (Figure A.3) is given in Appendix A.3. The base for the modular backward search is a partially composed model. The module structure below the level of distribution of the controller is dissolved by stepwise vertical composition (see Section 2.3). The result is a network of base modules. The composed module of the ejection unit is shown in Figure 6.5.

Controllability and observability of plant nodes (see Section 4.3) play a major role in the synthesis algorithm. The set of controllable transitions of the ejection unit is $T^c = \{t32, t33\}$. The set of controlled places to those transitions consequentially is $P^c = \{ev\_on, ev\_off\}$. Observable are the sensor places $P^o = \{2B1on, 2B1off\}$. The

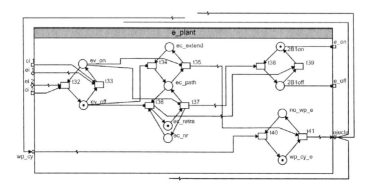

Figure 6.5: Partially composed $_SNCES$ model of the ejection unit.

list of all controllable, controlled and observable nodes is given in Table A.1.

The event synchronizations and p- invariants are determined. The determined state invariants for the lifting, the ejection and the measuring unit are given in Table 6.1. More details about state invariants are given in Section 2.4. A simplified notation is used for the invariants. For the given sets of places (Table 6.1) holds that the sum of markings always is one.

Next, the specifications are distributed to local state predicates extended by state atoms and communication predicates. A local state predicate consists of all state atoms referencing to places located in the same module (unit). Local state predicates resulting from the same global predicate are bound by communication variables. This leads to two local state predicates for $f3$:

$$(ZP = ec\_nr, CO = com_1^+) \text{ and } (ZP = mc\_extend, CO = com_1^-).$$

and extended by an event synchronization:

$$(ZP = ec\_nr \wedge \overline{2B1on}, CO = com_1^+) \text{ and } (ZP = mc\_extend, CO = com_1^-).$$

The assignment of a direction $+/-$ is user-defined only within the specification distribution and does not influence the backward search results. The forbidden state $f4$ is distributed to:

$$(ZP = ec\_nr \wedge \overline{2B1on}, CO = com_2^+) \text{ and } (ZP = lc\_path, CO = com_2^-).$$

Table 6.1: State invariants of the ejection and measuring unit models.

lifting unit	ejection unit	measuring unit
p-invariants		
$v\_up, uneutral;$	$ev\_on, ev\_off;$	$mv\_on, mv\_off;$
$dneutral, v\_down;$	$ec\_extend, ec\_path, ec\_retra;$	$no\_wo\_u, wp\_cyl\_u;$
$lc\_up, lc\_path, lc\_down;$	$ec\_retra, ec\_nr;$	$mc\_extend, mc\_ne;$
$wp\_up, wp\_path, wp\_down,$	$2B1on, 2B1off;$	$high, low, no\_height;$
$no\_wp;$		
$1B1on, 1B1off;$	no_wp_e, wp_cy_e	$3B1on, 3B1off;$
$1B2on, 1B2off$		$hs\_on, hs\_off;$
event synchronizations		
$lc\_up, 1B1off$	$ec\_retra, 2B1off$	$mc\_ne, 3B1on$
$lc\_down, 1B2off$	$ec\_nr, 2B1on$	$mc\_extend, 3B1off$
extension of p-invariants by the event synchronizations		
$\overline{1B1off}, lc\_path, \overline{1B2off}$	$ec\_extend, ec\_path, \overline{2B1off}$	

If a local state predicate of a set of local specifications (ZP, CO) equals, then the local specifications with equal predicate can be concentrated to one. Therefore, the communication predicates CO are combined disjunctively, as described for equal local state predicates within the backward search. The specifications related to the ejection unit result to:

$$(ZP = ec\_nr \wedge \overline{2B1on}, CO = com_1^+ \vee com_2^+).$$

The main step of the modular synthesis, the modular backward search, starts from the result of those preliminary steps. There is one local backward graph per unit module. The backward search is discussed more detailed for the ejection unit. The complete results of the example are in summary shown in Figure 6.6. The extension of the state predicates using the p-invariants is not included through the whole example because the state predicates would get confusingly large.

The first node of the graph of the ejection unit is

$$CP_{e,1} = \left(1, \left(ec\_nr \wedge \overline{2B1on}\right), com_1^+ \vee com_2^+\right).$$

The possible steps leading to a marking satisfying $ec\_nr \wedge \overline{2B1on} = 1$ are $\xi_1 = \{t37\}$ and $\xi_2 = \{t37, t38\}$. The transition $t37$ is enabled only if it is marking enabled by a token on the place ec_retra and condition enabled by a token on ev_on. Following the re-determination rule and the propagation rule, the pre-predicate ZP' results to $ZP' = ec\_retra \land ev\_on \land \overline{2B1on}$. It is extended using the event synchronization $ec\_retra + 2B1off = 1$ to the pre-predicate $ZP' = ec\_retra \land ev\_on \land \overline{2B1off} \land \overline{2B1on}$ containing a contradiction to the p-invariant $2B1off + 2B1on = 1$. Hence, the only valid pre-predicate is the one to ξ_2 with $ZP' = ec\_retra \land ev\_on \land 2B1on$. The communication predicate is inherited of $CP_{e,1}$ without extension because ξ_2 is a local possible step and also the possible step. Therefore, a new node $\left(2, (ec\_retra \land ev\_on \land 2B1on), com_1^+ \lor com_2^+\right)$ is included in the graph $BG_{e,1}$ together with the arc $BA_{e,1} = (ec\_nr \land \overline{2B1on}, \{t37\}, ec\_retra \land 2B1on \land ev\_on)$. The transitions $t37, t38$ are not controllable and no decision predicate is formulated. The predicate ZP' is treated as a new forbidden state predicate.

The backward search proceeds with a new run and the predicate $ZP = ec\_retra \land ev\_on \land 2B1on$. There are three possible steps $\xi_3 = \{t36\}, \xi_4 = \{t36, t39\}, \xi_5 = \{t33\}$ possibly causing a marking change from $ZP'(m(\mathcal{M}_e)) = 0$ to $ZP(m(\mathcal{M}_e)) = 1$. The pre-predicates are $ZP_3' = ec\_path \land ev\_on \land ec\_nr \land 2B1on$ for ξ_3, $ZP_4' = ec\_path \land ev\_on \land ec\_nr \land 2B1off$ for ξ_4 and $ZP_5' = ec\_retra \land ev\_off \land 2B1on$ for the possible step ξ_5. The pre-predicates ZP_3' and ZP_4' are covered by the state predicate of $CP_{e,1}$ and the CO is again inherited without change and identical to the one at the first node. Hence, both predicates are not treated further on. Beyond that, the extension of ZP_3' with the event synchronization would lead to the conflict predicate $ec\_path \land ev\_on \land ec\_nr \land 2B1on \land \overline{2B1on}$. The only valid pre-predicate (following Definition 7.1.5) is ZP_4', and the resulting node is $\left(3, (ec\_retra \land ev\_off \land \land 2B1on), com_1^+ \lor com_2^+\right)$. The transition $t33$ is controllable, and the built decision predicate is

$$DP = \left(\left(3, (ec\_retra \land ev\_off \land 2B1on), com_1^+ \lor com_2^+\right), cie1\right).$$

The local backward search now terminates because no untreated forbidden state predicate is left. The final local backward graph is shown in Figure 6.6 together with the graphs of both other modules.

The predicates of the nodes of the backward graph contain references to non-observable places. Only state atoms to observable places may be included in control functions. Hence, all predicates within the local backward graphs and decision predicates are converted to observable predicates including only references to place in $\{P^o \cup P^c\}$. The only decision predicate of the backward search graph of the ejection unit contains the observable place $2B1on$ and further the controlled and observable place ev_off. Hence, no further backward search for "real" observable predicates (see Algorithm 4) is indicated. The resulting

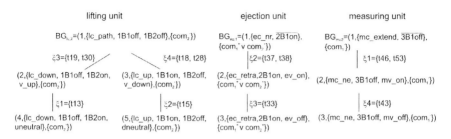

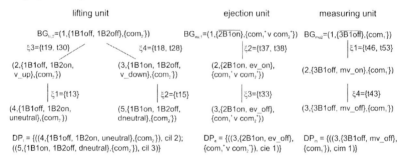

Figure 6.6: Resulting local backward graphs for the forbidden states $f3$ and $f4$ (extract of Figure A.1 in the appendix).

decision predicate is

$$DP = \left(\left(3, \left(ev\_off \wedge 2B1on \right), com_1^+ \vee com_2^+ \right), cie1 \right).$$

The local backward graphs and decision functions containing only observable state atoms are displayed in Figure 6.7 for all three units.

Figure 6.7: Local backward graphs for the forbidden states $f3$ and $f4$ containing observable state atoms only.

Decision functions are generated from the DP depending on the same condition input. Therefore, the state predicates are combined conjunctively with the communication predicates of the same node and converted to a DNF. If more than one node is element of a DP, then the DNFs of all node predicates are combined disjunctively. The decision function for the ejection unit is

$$cie1 \text{ is zero if } ev\_off \wedge 2B1on \wedge com_1^+ \vee ev\_off \wedge 2B1on \wedge com_2^+ = 1.$$

The communication functions of variable $com_x^{(s)}$ are composed of all nodes containing the variable with opposite $s = (+,-)$. That leads to the communication functions

$$com_1^- := 1 \text{ if } \overline{2B1on} \wedge com_2^+ \vee 2B1on \wedge ev\_on \wedge com_2^+,$$

$$com_2^- := 1 \text{ if } \overline{2B1on} \wedge com_1^+ \vee 2B1on \wedge ev\_on \wedge com_1^+.$$

The first conjunctive term of the functions results from the first node of $BG_{eu,1}$ and the second from the second node, respectively. Every control function results in the DNF and can by reduced using methods known for Boolean functions. All control functions of the whole example are arranged in Table A.2. All nodes of the local backward graphs contain at least one observable state atom and no predicate results to a simple true, therefore. If a node with an unobservable state predicate contains a communication variable com_x^s in CO, then the value of the opposite variable com^{-s} would result to always true. It is replaced by true within all decision functions it is contained within.

6.4 Performance of the modular backward search

The backward search as presented in this work produces useful results under incomplete state observation only with the use of event synchronizations. The backward search results without extension by event synchronizations would contain no sensor states. The alternative local backward graph is displayed in Figure A.1 as resulting from modular backward search without extension by state predicates. An analysis of the control functions shows that every communication functions would result to be always true. In such a case, a synthesis result would exist referencing to actuator values only. It would be much more restrictive than the result presented in the last paragraphs.

The presented approach has one drawback beside the major advantages in comparison with other approaches (also other backward search based approaches) summarized in Section 1.2. The Algorithm 2 is not maximally permissive because so-called forward effects of possible steps are not regarded. An efficient approach determining the forward effect of possible steps without neutralization of the advantages of modular analysis is left for later works. A maximally permissive variant of backward search is described in the following chapter. It is based on a composed plant bahavior model.

The modular backward search approach avoids the computation of the whole reachability space. A comparison of the number of computed state predicates (11 state predicates displayed in Figure 6.6) with the number of reachable states (6192 states calculated for the fully composed model of the testing station model as displayed in Figure A.3) shows a drastic reduction of computational complexity. State predicates cover a number of markings with a specific relevant property and avoid the handling of all complete markings with that property. The modular backward search also reduces the complexity compared to the monolithic backward search. The results of the monolithic approach introduced in [MH06b] are shown in Figure A.2. It consists of 21 determined predicates instead of 11 with the modular approach. The local predicates are determined just once independently of their global context in the modular approach. That context is handled in the communication predicates. In contrast, it can be seen that different global predicates determined by the monolithic approach include the same local predicates.

The presented approach is the first modular synthesis approach for distributed safety controllers considering incomplete state observation based on $_sNCES$ plant models. The modular procedure ensures that the analysis only extends over the model parts relevant under the given specification. Most steps of the synthesis excute with local plant models defined by the aimed distribution of the control. The only exception is the determination of possible steps as composition of local possible steps. Nevertheless, the approach particularly offers the possibility for parallel execution of the local backward search on the model of related plant parts combined with the possible step determination of the complete system model.

The control functions resulting from the presented modular synthesis approach can be transformed directly to executable controller program code. Transformation rules have to be worked out for the different kinds of program code, for example to IEC 61131 or IEC 61499 conform languages. Examples of such transformation and required transformation rules are presented in Chapter 8 for the generation of IEC 61499 conform Basic Function Blocks.

7 Synthesis of Forcing/Locking Safety Controllers

The method of backward search to define uncontrollable trajectories to forbidden states is introduced in the previous chapter. The existence of an uncontrollable trajectory from a state to a forbidden state does not imply all trajectories to lead to a forbidden state from this state. But the systems behavior can not be restricted to such an admissible trajectory by locking of any input. Otherwise, the trajectory wouldn't be called uncontrollable. Locking system's behavior via condition inputs is not the only possible control action. The event inputs of the plant allow to force the firing of steps. So, the forcing action of the safety controller can force the behavior to a trajectory not leading to a forbidden state. That is symbolically shown for the uncontrolled pre-region in Figure 7.1. It shows the alternative of forcing action compared to Figure 6.1.

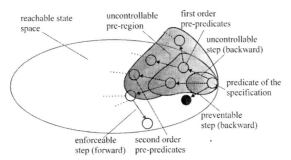

Figure 7.1: The uncontrollable pre-region for results with forcing and locking control action.

A first approach on forcing control function was presented in [LHR96]. The approach introduced in the following combines that idea with a maximally permissive backward search algorithm and an improved method to avoid cyclic control action. Further, a distribution method is given to generate distributed controller functions.

7.1 Synthesis of forcing/locking control functions

The synthesis approach for forcing/locking safety controllers introduced in the following uses two state analyzing procedures. The uncontrollable pre-region to forbidden states is explored by an improved monolithic backward search. It is called monolithic because it uses a fully composed plant model and results in monolithic control functions in contrast to the modular approach. The monolithic backward search was introduced in [MH06b] and is now enhanced to produce maximally permissive results. Passive or locking control components are derived by the backward search results. In addition, to every pre-predicate of a backward step, a forward analysis is used to find options to force the plant to avoid forbidden states. A forced step avoids forbidden states if:

- it leads to states not covered by the set of forbidden states in all step configurations,

- it does not lead into cycles of forced steps and

- it does not include a locked controllable transition.

Active control functions are derived from the forward analysis results. The passive control functions lock the state transitions by setting a condition input to zero. The forcing of state transitions is reached by activating an event input to controllable transitions. The method uses a fully composed plant model and global specifications. The resulting control functions are distributed to local control functions in an additional step. The controllability and observability of plant model elements is determined as defined in Section 4.3. The affiliation information of net elements to modules has to be stored before the composition for later distribution of the control functions.

7.1.1 Partially enabled controllable steps

A definition of forward analyzed steps under consideration of partial markings is needed to use the abstraction in terms of predicates. The possible steps are used as defined in Section 6.1. The forward step analysis permits the consideration of possible avoidance of forbidden behavior by forcing control interventions. Therefore, the definition of enabled steps is brought forward to partial markings defined by state predicates. The property of being not disabled by a predicate is defined for transitions as first step.

Definition 7.1.1. (Not Disabled Transitions)
A transition $t \in T$ of an $_sNCEM$ is:

1. *not marking disabled* under consideration of a state predicate ZP iff
   $(\not\exists p \in P$ with $(p,t) \in F : (m(p) = 0) \in ZP) \wedge$
   $(\not\exists p \in P$ with $(t,p) \in F : (m(p) = 1) \in ZP)$,

2. *not condition disabled* under consideration of a state predicate ZP iff
   $\not\exists p \in P$ with $(p,t) \in CN : (m(p) = 0) \in ZP$.

$\square$

Theorem 6. *Any under a marking $m(M) : ZP(m(M)) = true$ enabled set of transitions ξ (Definition 2.2.8) satisfies Definition 7.1.1. under the assumption $\forall c^{in} : is(c^{in}) = 1$.*

The default condition input state of an uncontrolled plant model is assumed as $is(c^{in}) = 1$. Only locking control functions influence the condition input state.

Proof. From $\forall p \in P$ with $(p,t) \in F : m(p) = 1 \equiv \not\exists p \in P$ with $(p,t) \in F : m(p) = 0$, $\forall p \in P$ with $(t,p) \in F : m(p) = 0 \equiv \not\exists p \in P$ with $(t,p) \in F : m(p) = 1$ and $\not\exists p \in P$ with $(p,t) \in CN : m(p) = 0 \equiv \forall p \in P$ with $(p,t) \in CN : m(p) = 1$ follows under $m(M) : ZP(m(M)) = true$ directly from Theorem 6. $\square$

Theorem 7. *The inverse of Theorem 6 only holds:*

- *for non-blocking $_sNCES$ if for all not disabled transitions t holds $\forall p \in P$ with $(p,t) \in F : \exists ZA = (m(p) = a) \in ZP$,*

- *for general $_sNCES$ additionally has to hold $\forall p \in P$ with $(t,p) \in F : \exists ZA = (m(p) = a) \in ZP$.*

Proof. The proof follows from the proof of Theorem 6 directly. $\square$

The model net is assumed to be non-blocking in the following.

Controllable steps influencing the marking of a state predicate ZP are considered only within the later described forward analysis. Controllable transitions are specified in Equation 4.3. *Partially enabled controllable steps* under consideration of state predicate ZP are defined referring to the set of controllable transitions as follows:

Definition 7.1.2. (Partially Enabled Controllable Step in non-blocking $_sNCES$)
Let $\mathcal{M}$ be an $_sNCEM$, ZP a state predicate, and $\widehat{\xi} \subseteq T$ a nonempty set of transitions within $\mathcal{M}$.
$\widehat{\xi}$ is a *partially enabled controllable step* iff

- it is a step (Definition 2.2.8) with $|T^c \cap \widehat{\xi}| = 1$,

- $\forall t \in \widehat{\xi}$ holds $\forall p \in P$ with $(p,t) \in F : \exists ZA = (m(p) = a) \in ZP$ and

- $\forall t \in \widehat{\xi}$ holds that they are *not marking disabled* and *not condition disabled*.

$\square$

All under a marking $m(M) : ZP(m(M)) = true$ enabled steps with $t_e \in T^c$ are also partially enabled controllable steps. That follows from Theorem 6 directly.

There can be more than one partially enabled controllable step, and the set of steps $\widehat{\xi}$ to a state predicate ZP, trigged by the same controllable transition t^c is $\widehat{\Xi}_{t^c,ZP}$. A maximal step cannot be defined because of consideration of only partial markings.

7.1.2 Monolithic backward search

The algorithm described in the following again uses the backward search to find all uncontrollable trajectories leading to a forbidden state. Thereby, lockable and enforceable controllable steps are determined. The main backward search element of the algorithm is described in the following and the additional forward analysis in the next section.

The backward search is processed on a fully composed plant model. The affiliation information of net elements to modules has to be stored before composition for later distribution of the control functions. The set of places is stored for every module on the distribution level. Afterwards, the whole plant model, which normally is a composite module, is vertically and horizontally composed as defined in the Definitions 2.3.1, 2.3.2 and 2.3.3. The result is a basic module with some open inputs and outputs. The composed model and forbidden state specifications in terms of state predicates or a state attribute are the starting point of the backward search. Any rules and definitions different to the ones in Section 6.2 are introduced in the following.

Some notations are defined to later describe the enhanced maximally permissive backward search. First, invalid transitions are defined under consideration of a possible step ξ_{ZP}^p and a related state predicate ZP.

Definition 7.1.3. (**Invalid Transitions**)

Invalid transitions $t^i_{\xi^p;ZP} \in T$ are transitions for which holds:

- $t \notin \xi^p_{ZP}$ and t is not in conflict with a transition $t' \in \xi^p_{ZP}$,

- $\exists p \in P : (p,t) \in F|(m(p) = 1) \in ZP \vee \exists p \in P : (t,p) \in F|(m(p) = 0) \in ZP$,

- $\nexists p \in P : (p,t) \in F|(m(p) = 0) \in ZP \wedge \nexists p \in P : (t,p) \in F|(m(p) = 1) \in ZP$ and

- $\nexists p \in P : (p,t) \in CN|(m(p) = 0) \in ZP \wedge \nexists(t',p) \in F : t' \in \xi^p_{ZP}$

$T^i_{xi^p;ZP}$ is the set of all invalid transitions depending on a state predicate ZP and a possible step ξ^p.

□

Such invalid transitions are not within a possible step by definition. Possible steps are not maximal and an extension of them by additional event signal sink transitions could include invalid transitions. Such a transition would remove a token from places contained in $ZA \in ZP$ or mark places $\overline{ZA} \in ZP$ if it was part of an enabled step ξ including the analyzed possible step $\xi^p_{ZP} \subseteq \xi$. Moreover, it is not disabled under a state satisfying ZP. An example with an invalid transition is shown in Figure 7.2. The forbidden state predicate is $\{p2 \wedge p3\}$, symbolized by the double circuits. An extension of the possible step $\{t1\}$ to an enabled step $\{t1, t2, t3\}$ would not lead to a state satisfying the forbidden predicate $\{p2 \wedge p3\}$. Transition $t3$ is invalid under consideration of the condition in Definition 7.1.3.

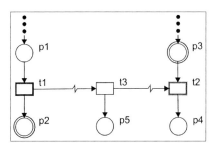

Figure 7.2: Example of an invalid transition $t2$ to a possible step $\{t1\}$ and a predicate $p2 \wedge p3$.

Proposition 7.1.1. *Steps including invalid transitions cannot lead to a follower marking* $m'(\mathcal{M})$ *satisfying* $ZP(m'(\mathcal{M})) = 1$, *i.e.* $\nexists t^i_{xi^p;ZP} \in T^i_{xi^p;ZP}|t^i_{xi^p;ZP} \subset \xi : ZP(m'(\mathcal{M})) = 1$.

Proof. The first, third and fourth condition of the definition of invalid transitions ensure that a transition t^i is enabled (Definition 2.2.3) under every state satisfying ZP. Consequent from Definition 2.2.9 holds for all transition within an enabled step, so also for a t^i:

$\forall p \in P : \exists (t,p) \in F \wedge t \in \xi$ holds $m'(p) = 1$ and

$\forall p \in P : \exists (p,t) \in F \wedge t \in \xi$ holds $m'(p) = 0$.

For invalid transitions $t^i_{xi^p;ZP}$ holds $\exists p \in P : (p,t) \in F | (m(p) = 1) \in ZP \vee \exists p \in P : (t,p) \in F | (m(p) = 0) \in ZP$, i.e. the marking of a pre-place is relevant for $ZP(m(\mathcal{M})) = 1$

$\longrightarrow \forall \xi \cap T^i_{xi^p;ZP} \neq \emptyset : ZP(m') = 0$ $\square$

Next, *minimal possible step extension paths* are defined as a set of transitions being possibly part of an enabled step and including a possible step along with an invalid transition.

Definition 7.1.4. Minimal Possible Step Extension Paths

A *minimal possible step extension path* ξ^{pe} to a possible step ξ^p is set of a transition within $\mathcal{M}$ for which holds:

- $\exists t \in \xi^{pe} : \exists (t',t) \in EN \wedge t' \in \xi^p$,

- $\left| \xi^{pe} \cap T^i_{xi^p;ZP} \right| = 1$,

- $\{\xi^p \cup \xi^{pe}\}$ is a step (Definition 2.2.8) and

- $\nexists t \in \xi^{pe} | t \notin T^i_{xi^p;ZP}$ for which holds

  - $\exists p \in P : (p,t) \in F | (m(p) = 1) \in ZP \vee \exists p \in P : (t,p) \in F | (m(p) = 0) \in ZP$,

  - $\exists p \in P : (p,t) \in CN | (m(p) = 0) \in ZP \wedge \nexists (t,p) \in F : t \in \xi^p_{ZP}$,

  - $\exists t \in \xi^{pe} : \xi^{pe} - t = \xi'^{pe}$.

Ξ^{pe} is the set of all minimal possible step extension paths to one ξ^p.

                                                                                                   $\square$

Minimal possible step extension paths are paths of event signal connected enabled transitions, which are not part of the analyzed possible step but which possibly extend that step to include invalid transitions. The path is minimal means thereby, that deleting one transition would lead to a transition set not satisfying the conditions of Definition 7.1.4. Enabled steps including the transitions of ξ^p_{ZP} and ξ^{pe}_{ZP} cannot lead to marking satisfying ZP. This follows from the enclosure of an invalid transition in such step in accordance with Proposition 7.1.1. For the example given in Figure 7.2 ξ^{pe} is the transition set $\{t2,t3\}$.

Based on those extensions, the backward search is defined similarly to Section 6.2 but not in a modular way. Instead, the results will be maximally permissive and give the basis for forcing control functions determination. The *re-determination rule*, the *propagation rule* and the *consistence property* are used as defined in Section 6.2. They are completed by the following introduced *forward analysing rule*. It is introduced to consider so-called forward effects of possible steps. An enabled step including a possible step ξ^p can also include invalid transitions. Such step does not lead into a forbidden marking (Proposition 7.1.1) and should not be prevented by a controller intervention, therefore.

There will exists a forward effect if a minimal possible step extension path (see Definition 7.1.4) to a possible step exists.

Only pre-predicates surely not enabling steps with forward effects are further considered in the control function synthesis because of the aim to generate a maximally permissive controller. In other words, only enabled steps to states subsumed by pre-predicates disabling such an extension path can lead to a forbidden marking and have to be taken into account. For every ξ^{pe} has to hold that $\exists t \in \xi^{pe}$, which is disabled by ZP'. This ensures capturing all enabled steps that do not include the invalid transition. All possible configurations of disabling have to be taken into account based on the rules for marking and condition enabling in Definition 2.2.3.

Therefore, the following *forward analyzing rule* is defined:
State predicates are added to a pre-predicate's set $\mathbb{ZP}'$ for every possible step extension path according to the following definitions:
$\forall p \in P : (p,t) \in F \wedge (m(p) = 1) \notin ZP \wedge t \in \xi^{pe} \Rightarrow ZP' \wedge \overline{ZA'} = (m(p) = 0) \in \mathbb{ZP}'$ and
$\forall p \in P : (t,p) \in F \wedge (m(p) = 0) \notin ZP \wedge t \in \xi^{pe} \Rightarrow ZP' \wedge ZA' = (m(p) = 1) \in \mathbb{ZP}'$ and
$\forall p \in P : (p,t) \in CN \wedge (m(p) = 1) \notin ZP \wedge t \in \xi^{pe} \Rightarrow ZP' \wedge \overline{ZA'} = (m(p) = 0) \in \mathbb{ZP}'$ and
if $\exists t \in \xi^{pe} : (c^{in}, t) \in CI^{arc}$, then $ZP' \cup \mathbb{ZP}'$ is generated applying the re-determination rule and the propagation rule to all transitions $t \in (\xi^p \cup \xi^{pe})$. The generated pre-predicates have to be part of the control components as defined later in this section.
All $ZP \in \mathbb{ZP}'$ have to be free of contradictions.

The original ZP' is not part of $\mathbb{ZP}'$ if there exists a forward effect. Following, the set $\mathbb{ZP}'$ can be empty. An empty $\mathbb{ZP}'$ means that the forward effect to ξ^p is included in any enabled step ξ with $\xi^p \in \xi$. The step therefore cannot lead into a forbidden state of ZP, and no control predicate or forbidden state predicate has to be defined. The specific state predicates of $\mathbb{ZP}'$ describe the disabling of just one transition of ξ^{pe}. All predicates with combinations of more than one of those transitions are covered by the generated predicates.

Proposition 7.1.2. *The set of state predicates $\mathbb{ZP}'$ determined by the forward analyzing rule covers all markings that enable a step extension of ξ^p not including any invalid transition.*

Proof. in two steps:

- All ξ^{pe} are minimal per definition. That means that by disabling one transition, the set cannot be part of an enabled step.

- Every added state predicate contradicts one condition of Definition 2.2.3 → the extended ZP' does not enable a step including the ξ^{pe}. The rule includes all places not covered by a state atom of ZP.

$\square$

The modeling experience according to well-structured models shows that the models of real systems mostly consist of rather short event graphs with three to five transitions that are connected by events. For such system models, the extended set of pre-predicates contains just a few predicates. The number of predicates depends on the number of pre- or post-places and condition signal sources to the transitions in ξ^{pe} that are not included in the state predicate ZP.

Invariants are checked to improve the efficiency of the backwards search and extend the resulting state predicates to improve the results under partial observation, as described in Section 6.2.

The set of pre-predicates $\mathbb{ZP}'$ is defined under consideration of the mentioned rules .

Definition 7.1.5. (Pre-Predicate)
Let $\mathcal{M}$ be an $_sNCEM$, ZP a state predicate and ξ^p a possible step over $\mathcal{M}$. The set of pre-predicates $PZP(ZP, \xi^p)$ of ZP respecting $\xi^p \in \Xi^p$ $(ZP'[\xi^p\rangle ZP)$ is the set of state predicates $ZP' \in \mathbb{ZP}'$, which is calculated by backward analyzing steps ξ^p under consideration of the re-determination rule, the propagation rule, the forward analyzing rule and the consistence property and for which the set of markings fulfilling a $ZP' \in \mathbb{ZP}'$ is not within the set of markings fulfilling ZP.

$\square$

A set of all determined forbidden states FSP (starting with $FSP = \mathbb{ZP}$) is built up during the backward search. It is used during the subsequent forward analysis and represents the uncontrollable pre-region of a forbidden state when the backward search has been terminated. If the firing of the analyzed possible step can be omitted by a control action, then a control component is generated. Else, the pre-predicates are forbidden too and are therefore origin of another backward analysis if no forcing action is possible. The enabling of steps can be omitted in two ways. First, a possible step can be locked directly if it contains a controllable transition. The generated control component has to set the condition input connected to the controllable transition to zero. Second, the forcing of an alternative step causes the immediate leaving of the states covered by ZP' and omits enabling of an enabled step containing the possible step. The necessary forward analysis is described in Section 7.1.4.

Algorithm 5 describes the backward search based on Definition 7.1.5. The gray parts of the algorithm are placeholders for the forward step analysis introduced in the next section. Skipping these parts leads to pure maximally permissive locking controller synthesis.

The control functions are composed from the elements of the set of Control Components CC after termination of the backward search. The pre-predicates related to the same condition input are combined to the control function CCF_i for the input c_i^{in} disjunctively. It is a function

$$ CCF_i = \bigvee ZP' : \forall ZP' : \exists (ZP', c_i^{in}) \in CC \wedge \nexists CCF_n \supseteq (ZP', c_i^{in}). $$

The input states of condition inputs of the plant are defined under evaluation of the control functions. The input state of all condition inputs is 1 or the value of the process controller output $c_{i,p}^{in}$ if it exists for the plant input, by default. For a controlled plant model with a marking $m(M)$ holds: $is(c_i^{in}) = \begin{cases} 0 & if \exists CCF_i : CCF_i(m(M)) = true \in F \\ c_{i,p}^{in} & if \exists c_{i,p}^{in} \in C_{i,p}^{in} \\ 1 & else. \end{cases}$

The result is a set of monolithic control functions for the (local) locking controller(s). Some properties of the backward search are discussed in the next section and it is proven to be maximally permissive.

7.1.3 Permissiveness of the backward search results

A major criterion for supervisor and safety controller synthesis results is maximally permissiveness. That means that the resulting controller components or functions avoid unsafe behavior only. Control components are maximally permissive if there exists no control of

while $FSP \neq \emptyset$ **do**

   foreach $ZP \in FSP$ **do**

      determine the set Ξ_{ZP}^p of all possible steps to the predicate ZP;

      foreach $\xi^p \in \Xi_{ZP}^p$ **do**

         determine the set Ξ^{pe} of all possible step extension;

         determine the set of pre-predicates $\mathbb{ZP}'$ enabling the possible step ξ^p under consideration of Ξ^{pe};

         check state invariants and extend all $ZP' \in \mathbb{ZP}'$ according to the state predicates;

         if ξ^p *is not controllable* **then**

            perform *forward analysis* (Algorithm 6);

            if ZP' *is a forcing state* **then**

               build forcing control component;

            else

               include $FSP' = FSP' \cup \mathbb{ZP}'$, $FSP = FSP/ZP$ and $AFS := AFS \cup \mathbb{ZP}'$.

            end

         else

            include ZP' and the controllable condition input $c^{in} : (c^{in}, t^c) \in CI^{arc} \wedge t^c \subseteq \xi^p$ in the set of control components $CC = CC \cup (ZP', c_i^{in})$ and $FSP = FSP/\mathbb{ZP}'$;

         end

      end

      validate all forcing states (Algorithm 7);

   end

   $FSP := FSP'$ and $FSP' := \emptyset$;

end

Algorithm 5: Backward search algorithm

the same type (in our case locking safety control) that ensures the same specification but is less restrictive.

Definition 7.1.6. Maximally Permissive Controller

A set of control components CC with a set of forbidden markings

$$M_f = \sum_{FSPcov\,m(p)} m(p)$$

is maximally permissive if no CC' exists for which holds

$$M'_f \subset M_f.$$

$\square$

In other words, a safety controller is maximally permissive if all states covered by the set of forbidden state predicates $ZP \in FSP$ have an uncontrollable trajectory to a forbidden state of the specification.

The synthesized controller has to satisfy the following conditions to avoid only those markings covered by a $ZP \in FSP$ that has an uncontrollable trajectory to a forbidden state of the specification:

- all prevented steps are uncontrollable

- all possible steps ξ^p depending on a pre-state that satisfies ZP' lead into a state satisfying a forbidden state ZP,
  - (while every pre-predicate to an uncontrollable possible step ξ^p to a forbidden state predicate is also a forbidden predicate),

- the last condition also holds for all extensions to enabled steps ξ depending on a state satisfying ZP' for which holds $\xi^p \subseteq \xi$.

The first condition is true by definition of possible steps and backward search.

Subsequent propositions are stated before the proof of the second condition.

Proposition 7.1.3. *Possible steps ξ^p_{ZP} to a predicate ZP are a set of enabled transitions (not enabled steps according to Definition 2.2.8 because they are extendable) under consideration of the partial marking defined by $ZP' \in PZP(ZP, \xi^p_{ZP})$ and ZP calculated according to Definition 7.1.5 and if $\forall (c^{in}, t) \in CI^{arc} : t \in \xi^p$ holds $is(c^{in}) = 1$.*

Proof. It holds from the re-determination rule that:

$$\forall p \in P : \exists t \in \xi^p : (p, t) \in F | m'(p) = 1 \land p \in ZP'$$

$$\forall p \in P : \exists t \in \xi^p : (t,p) \in F | m'(p) = 0 \wedge \overline{p} \in ZP'$$

$$\forall p \in P : \exists t \in \xi^p : (p,t) \in CN | m'(p) = 1 \wedge p \in ZP'$$

$\to \xi_{ZP}^p$ is marking and condition enabled according to Definition 2.2.3 if no control input c^{in} is set to zero.   □

Now, the second condition listed on the last page is handled in two steps as follows.

Theorem 8. *All ξ_{ZP}^p lead into a forbidden state ZP, $ZP'[\xi^p\rangle ZP$. At this, all state predicates ZP' are forbidden inductively for which exists an uncontrollable ξ_{ZP}^p.*

Proof. It is assumed that backward search starts from a forbidden state predicate. Then the proof follows the steps:

- $\forall t \in \xi_{ZP}^p$ holds they are enabled through a ZP' (Theorem 7.1.3).

- ξ^p consists of leastwise one transition t with $(t,p) \in F : (m(p) = 1) \in ZP$ or a transition t^* with $(p,t^*) \in F : (m(p) = 0) \in ZP$ by definition of possible steps.

- Resulting from Definition 2.2.9, a place $(m(p) = 1) \in ZP$ is marked by firing a transition t and a place $(m(p) = 0) \in ZP$ gets unmarked by firing a transition t^*. Considering the propagation rule holds
  $\{\forall p \in P : (m(p) = 1) \in (ZP' \cap ZP) \to \nexists(t,p) \in F : t \in \xi^p\} \wedge$
  $\{\forall p \in P : (m(p) = 0) \in (ZP' \cap ZP) \to \nexists(p,t^*) \in F : t^* \in \xi^p\}.$
  Together follows
  $\forall p \in P \wedge (m(p) = 1) \in ZP : ((m(p) = 1) \in ZP') \vee$
  $(\exists(t,p) \in F : t \in \xi^p \wedge (\exists(p',t) \in F : (m(p') = 1) \in ZP' \vee \nexists p' \in P : (p',t) \in F))$ and
  $\forall p \in P \wedge (m(p) = 0) \in ZP : ((m(p) = 0) \in ZP') \vee$
  $(\exists(p,t) \in F : t \in \xi^p \wedge (\exists(t,p'') \in F : (m(p'') = 0) \in ZP' \vee \nexists p'' \in P : (t,p'') \in F))$
  $\to ZP'[\xi^p\rangle m(N) : m(N) \, satisfies \, ZP.$

□

The next step is proving Theorem 8 to hold also for all enabled steps ξ for which $\xi_{ZP}^p \subseteq \xi$ holds. Such steps are denoted as ξ^{p+}.

Theorem 9. *All extensions of ξ^p that are not disabled by a $CC = (ZP', c^{in})$ lead into a forbidden state ZP ($\forall \xi^{p+} : ZP'[\xi^{p+}\rangle ZP$). In this again, all state predicates ZP' are inductively forbidden for which exists an uncontrollable ξ_{ZP}^p.*

Proof. It holds for all $t \in T : (p, t) \in F \land (m(p) = 1) \in ZP$ that they are invalid or disabled by ZP or cannot form a step $\xi \supseteq \xi_{ZP}^p$ (follows from Definition 7.1.3).

It holds for all $t \in T : (t, p) \in F \land (m(p) = 0) \in ZP$ that they are invalid or disabled by ZP or cannot form a step $\xi \supseteq \xi_{ZP}^p$ (follows from Definition 7.1.3)

It holds for all $t \in T : (p, t) \in F \land (m(p) = 0) \in ZP$ that they are part of a possible step (Definition 6.1.2) or cannot form a step (following Theorem 1).

It holds for all $t \in T : (t, p) \in F \land (m(p) = 1) \in ZP$ that they are part of a possible step (Definition 6.1.2) or cannot form a step (following Theorem 1).

All invalid transitions are disabled by ZP' and must not be part of an enabled step as a consequence of Proposition 7.1.2. $\implies$ all extensions to ξ_{ZP}^p, which can form a step (Definition 2.2.8) lead to a forbidden state predicate ZP. $\square$

Theorem 10. *The set of Control Components CC generated by monolithic backward search as described in Section 7.1.2 is maximally permissive.*

Proof. From Theorems 8 and 9 it follows that at least one enabled step not leading to a forbidden state ZP exists for states satisfying a forbidden state ZP'. Hence, every pre-predicate has a trajectory to a forbidden state of the specification. Control functions CC consist of pre-predicates to controllable steps ξ_{ZP}^{pc} and are not declared to be forbidden state predicates by definition. Therefore, for all markings m with $ZP'(m)$ holds there is at least one uncontrollable step $|\xi \cup T^c| = 0$ to a marking covered by a forbidden state predicate $\forall ZP' \in FSP : \exists \xi \in \Xi : ZP'[\xi] ZP \land ZP \in FSP$. From this follows that the set of forbidden markings M_f is minimal and no more permissive controller exist. $\square$

The locking control predicates determine the last states before entering the uncontrollable region to forbidden states of a locking controller. The following defined forward step analysis builds up on that to result in forcing/locking control functions.

7.1.4 Forward step analysis

Forward steps and their follower markings to all forbidden pre-predicates[1] of a backward search step are analyzed. The aim is to find forceable state transitions avoiding any uncontrollable trajectory to forbidden states.

---
[1] pre-predicates to uncontrollable possible steps

Similar to the backward search, the follower predicates ZP^* and a set of partially enabled controllable steps $\widehat{\Xi}_{t^c,ZP}$ are defined to every analyzed state predicate ZP. The following rule is defined therefore.

The *forward determination rule* follows from Definition 2.2.9 with:

- $\forall p \in P : \exists t \in \widehat{\xi} : (p,t) \in F \Rightarrow ZA^* = (m'(p) = 0) \in ZP^*$, notation $\overline{ZA^*}$,
- $\forall p \in P : \exists t \in \widehat{\xi} : (t,p) \in F \Rightarrow ZA^* = (m'(p) = 1) \in ZP^*$,
- $\forall p \in P : \exists ZA = (m(p) = a) \in ZP \wedge \nexists t \in T$ with $(p,t) \in F : t \in \widehat{\xi} \wedge \nexists t \in T$ with $(t,p) \in F : t \in \widehat{\xi} \Rightarrow ZA = (m(p) = a) \in ZP^*$.

The resulting predicate ZP^* is considered only if it is free of contradictions with state invariants. Further, every predicate ZP^* has to be free of contradictions. Thus, ZP^* has to satisfy the *consistence property* defined in Section 6.2.

The follower predicates ZP^* are determined under consideration of the mentioned rules.

Definition 7.1.7. Follower Predicate
Let M be an $_SNCEM$, ZP a state predicate and $\widehat{\Xi}_{t^c,ZP}$ a set of partially enabled steps to ZP with the controllable transition t^c. The set of follower predicates $FZP(ZP, \widehat{\Xi}_{t^c,ZP})$ of ZP respecting $\widehat{\xi} \in \widehat{\Xi}_{t^c,ZP}$ is the set of state predicates ZP^*, which is calculated by forward analyzing steps $\widehat{\xi}$ under consideration of the forward determination rule, the consistence property and for which the set of markings fulfilling ZP^* is not covered (see Definition 5.1.3) by ZP.

$\square$

A forcing control intervention can avoid entering an uncontrollable trajectory to forbidden states only if the follower markings of the trigger step are not forbidden themselves. Further, the follower markings are not allowed to cause cyclic forcing control intervention. Cyclic forcing interventions trap the controller and so the plant behavior in a nonterminating cycle.

Definition 7.1.8. Forcing Predicate
A predicate ZP is a forcing predicate FP if a set of follower predicates $FZP(ZP, \widehat{\Xi}_{t^c,ZP})$ exists for whose members $\forall ZP^* \in FZP$ holds that they are not covered by a predicate within the set of forbidden states FSP and furthermore adding ZP to the set of forcing predicates $\mathbb{FP}$ is not causing the existence of a sequence $[FP_0, \widehat{\xi}_{t^c,FP_0}, FP_1, \ldots, FPk, \widehat{\xi}_{t^c FP_0}]$ i.e. a cycle of forcing predicates and forced steps.

$\square$

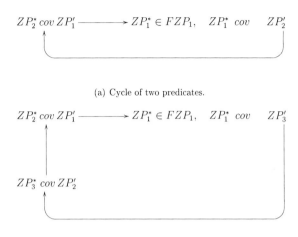

(a) Cycle of two predicates.

(b) Cycle of three predicates, while $ZP'_1, ZP'_2 \in ccFCC$.

Figure 7.3: Scheme of the two types of cycles of forcing states.

Algorithm 6 describes the forward analysis to be embedded in Algorithm 5. It results in a set of *Forcing Control Components FCC*. An additional set of cycle candidates FCCs (shortened $ccFCC$) is introduced to avoid unnecessary checks of new state predicates for possible initiation of a cycle of forcing control action. The set $ccFCC$ is built of all forcing predicates and their successors, which have a successor covering another predicate in the set of forcing predicate FCC.

Only the subset $ccFCC$ of FCC is checked, which contains ZPs having a successor ZP^* that covers another predicate of FCC. The check of an ZP^* to be added to FCC on closing a cycle will be performed only if the covered predicate is within $ccFCC$ or the set $ccFCC$ is empty and a predicate in FCC is covered by a successor of the analyzed predicate ZP'. Inductively, forcing predicates without a successor in $ccFCC$ cannot cause cyclic forcing. That is shown with the two minimal development examples of a cycle in Figure 7.3. Figure 7.3(a) shows the first variant of a cycle, which is the only case without covering a predicate in $ccFCC$. The second case is the minimal cycle with covering of predicate(s) in $ccFCC$ and shown in Figure 7.3(b).

The partial enabled steps are forced through setting the plant input interconnected with t^c of the step.

forward analysis;

determine the set of controllable transitions t^c under consideration of ZP';

foreach $t^c \in T^c$ **do**

  determine all *partial enabled step* configurations $\widehat{\Xi}_{t^c, ZP'}$;

  foreach $\widehat{\xi} \in \widehat{\Xi}_{t^c, ZP'}$ **do**

    determine the follower predicate ZP^* and include ZP^* in FZP;

  end

  if $\forall ZP^* \in FZP$ *holds* $ZP^* \notin FSP$ **then**

    $cycle(FZP) := false$;

    foreach $ZP^* \in FZP$ **do**

      `// check if adding `ZP'` in `FSP` results in cyclic forcing`

      if $\exists ZP_1 \in FCC : ZP^* cov ZP_1$ **then**

        if $\exists ZP_2 \in FZP(\Xi, ZP_1) : ZP_2 cov ZP'$ **then**

          $cycle(FZP) := true$;

        else

          if $\exists ZP_3 \in ccFCC : ZP^* cov ZP_3 \wedge ZP_4 \in ccFCC : ZP_4 cov ZP'$ **then**

            if $ZP' \cup ccFCC$ *contains a sequence*

            $[ZP_0, \widehat{\xi}_{t^c, ZP_0}, ZP_1, \ldots, ZP_k, \widehat{\xi}_{t^c FP_k}, ZP_0]$ **then**

              $cycle(FZP) := true$;

            end

          end

        end

        if $cycle(FZP) = false$ **then**

          $ccFCC := ccFCC \cup ZP' \cup ZP_1$;

        end

      end

    end

    if $cycle(FZP) = false$ **then**

      ZP' is a *forcing state*;`// evaluated in Alg. 5`

      $T^c := \emptyset$;

    else

      $T^c := T^c / t^c$;

    end

  else

    $T^c := T^c / t^c$;

  end

end

Algorithm 6: Forward analysis algorithm.

Theorem 11. *Triggering a step by* $is(e^{in}) = 1$ *with* $(e^{in}, t^c) \in EN^{in} \wedge FZP(ZP', \widehat{\Xi}_{t^c, ZP'})$ *at a marking* $m(M) : ZP'(m(M)) = 1 \wedge ZP' \in FP$ *avoids entering forbidden states, while* ZP' *is a pre-predicate to an uncontrollable forbidden trajectory.*

Proposition 7.1.4. *Backward search determines uncontrollable trajectories to forbidden states (follows from Theorem 8).*

Proof. From Definition 7.1.2 together with Theorem 6 follows that under a marking $m(\mathcal{M})$ with $ZP(m(M)) = 1$, the condition $ZP \in FP$, and set of under $m(\mathcal{M})$ enabled steps Ξ holds:

- $\exists \xi \in \Xi : \exists \xi_{t^c, ZP} \in \Xi_{t^c, ZP} : \xi_{t^c, ZP} \subseteq \xi \wedge \exists (e^{in}, t^c) \in EN^{in}$ and further

- $\forall \xi \in \Xi : \exists \xi_{t^c, ZP} \in \Xi_{t^c, ZP} | \xi_{t^c, ZP} \subseteq \xi \wedge \exists (e^{in}, t^c) \in EN^{in}$ holds that $\exists ZP^* \in FZP(ZP, \Xi_{t^c, ZP})$.

For all ZP^* it is true by Definition 7.1.8 that they are not part of an uncontrollable trajectory to forbidden states. Thus, the activation of the event input e^{in} avoids entering a forbidden state. □

If a predicate ZP' is determined as forcing predicate, then ZP' is added to the set of Forcing Control Components (FCC). The set is defined as $FCC = FCC \cup (ZP', e^{in}, FZP)$ with $(e^{in}, t^c) \in EN^{in} \wedge FZP(ZP', \widehat{\Xi}_{t^c, ZP'})$.

The inclusion of state predicates to FCC is not final before the backward search terminates because the backward search is performed step by step for more than one forbidden state. States, which have been defined being forcing states earlier, can become forbidden within the calculation process. These forbidden forcing states then have to be considered as forbidden states and deleted from FCC. They get part of the predicate set of forbidden state predicates FSP the backward search is performed for. The forcing states are checked for being forbidden or having forbidden successors depending on the newly calculated forbidden states after every calculation cycle (see Algortihm 5). The Algorithm 6 (called in Algorithm 5) describes the repeatingly proceeded validation of forcing states.

The check has not necessarily to be processed every cycle. It has to be completed with no changes as result before the control components are taken as final. The check can get complex for bigger controller models.

A forbidden forcing predicate is excluded from the forcing control components and $ccFCC$. If just a follower predicate of a forcing state is forbidden, alternative forceable transitions,

foreach $ZP \in FCC$ **do**

  if $ZP \in AFS$ **then**

   $FSP' := FSP' \cup ZP;$

   $FCC := FCC/FZP(ZP,\widehat{\Xi});$ $ccFCC := ccFCC/FZP(ZP,\widehat{\Xi});$

  else

   if $\exists ZP^* \in FZP(ZP,\widehat{\Xi}) : ZP^* \in AFS$ **then**

    perform *forward analysis* (Algorithm 6);

    if ZP' *is a forcing state* **then**

     build forcing control component;

    else

     $FSP' := FSP' \cup ZP;$

     $FCC := FCC/FZP(ZP,\widehat{\Xi});$ $ccFCC := ccFCC/FZP(ZP,\widehat{\Xi});$

    end

   end

  end

end

Algorithm 7: Validation of all forcing states.

their related partially enabled steps and follower predicates are determined. Only if no allowed alternative exists, then the forcing predicate is excluded too.

The termination criterion for the combination of backward search and forward analysis is that the set of forcing predicates contains no forbidden states or predicates with successors, which are covered by a forbidden state. Further, the set of forbidden states and uncontrollable possible steps has to be empty.

The monolithic forcing control functions (FCF) are derived from the FCC's according to Proposition 11 as $FCF_i = \bigvee ZP : \forall ZP : \exists (ZP, e_i^{in}) \in FCC \wedge \nexists CF_n \supseteq (ZP, e_i^{in})$.

As described before, the forward step analysis is fitted into the backward search algorithm given in Section 7.1.2 (gray parts in the algorithm). The combination of forcing and locking control interventions leads to control functions less or equal restrictive regarding to the admissible behavior compared to a simple maximally permissive locking controller. The forcing control interventions are related to states, which would be forbidden by such pure locking controller. That fact can be easily seen in the Algorithm 5 in Section 7.1.2. Following, every result including forcing states is less restrictive than its counterpart caused by just locking state transitions.

7.2 Distribution and example

The observation and control is structured in a way that it can be delegated to different controllers when designing distributed controllers. The synthesized monolithic control functions are distributed into local control functions for every distributed controller. Therefore, local predicates are calculated as defined in Definition 5.2.2. One controller is mapped to every plant unit. Thus, the observation of every distributed controller is one unit and it controls the actuators of this unit. The membership information of model elements stored before composition is used for this purpose.

The distributed controllers exchange state information about the local observation with other controllers. Figure 3.1 depicts the structure of the controller's interactions. The two kinds of resulting local control functions are similar to those introduced in Section 6.3. The major difference to the modular approach is that the communication functions only contain local observations and no communication variables. Decision functions are defined for event- and condition inputs of the plant for every local controller. The values of communication variables are used by other controllers for decision making. The local decision functions $DF^{M_x}_{c^{in},e^{in}}$ consist of local observations, i.e. observations of the controllers' related plant parts and of communication variables. The local decision functions $DF^{M_x}_{c^{in},e^{in}}$ are defined as

$$DF^{M_x}_{c^{in},e^{in}} = \bigvee_{\forall ZP_i \in CF_{c^{in},e^{in}}} \left(ZP^{M_x}_i \bigwedge_{\forall M_y \in N} CV^{M_y}_i \right), \qquad (7.1)$$

while

$$CV^{M_y}_j = | \bigwedge_{\forall ZA^{M_y} \in ZP^{M_y}_i} ZA^{M_y} |. \qquad (7.2)$$

The index j is used consecutively for all generated communication variables of the controller model. The values of event inputs are defined as

$$is(e^{in}_{M_x}) := DF^{M_x}_{e^{in}}.$$

The condition inputs have to be locked in all states satisfying the decision function, and that means formally

$$is(c^{in}_{M_x}) := !DF^{M_x}_{c^{in}}.$$

The values of the communication variables represent the observations of other controllers and are determined as

$$com_j := CV^{M_y}_j.$$

Of course, communication functions and variables are only generated if they do not exist in the set of already generated functions. The distribution of the functions leads to the information structure for the distributed controllers shown in Figure 3.1.

The synthesis of a distributed forcing/locking controller is exemplified by means of the lifting and the ejection unit of the example model of the testing station (Figure A.3) and the specification f1 and f4 as listed in Section 5.1. The forbidden states are:

f1 v_up   ∧   It must not be possible to enable the actuating valves for up and for
    v_down   down movement together.

f4 lc_path ∧   The ejection pusher is not allowed to move while the lifter is moving.
    ec_nr

The first predicate depends only on the lifting unit. The predicate $f4$ includes information of the lifting and the ejection unit. The forbidden state predicate $f4$ can be extended by an event synchronization to $lc\_path \land 1B1off \land 1B2off \land ec\_nr \land \overline{2B1on}$. A summary of all event synchronizations and p-invariants is given in Table 6.1.

All calculated predicates are also extended using the p-invariants. The extension is important to satisfy the forward step condition of having a set of pre-places with markings fully defined by state atoms (Definition 7.1.2).

The possible steps are analyzed starting from a specified forbidden state predicate. The possible steps leading to $v\_up \land v\_down$ are $\xi_1^p = \{t13\}$ and $\xi_2^p = \{t16\}$. The pre-predicates to those steps are determined next. All pre-places and source places of condition signals to step transitions are included in the predicates based on the enabling condition for steps. The post-places of step transitions are included with $ZA = m(p) = 0$. Both steps are controllable and therefore the extended pre-predicates are control components.

$$CC = ((uneutral \land \overline{v\_up} \land v\_down \land \overline{dneutral}); cil2;$$

$$(\overline{v\_up n neutral} \land \overline{v\_down} \land dneutral; cil3)).$$

The second forbidden state predicate is analyzed for possible steps next. The transitions $t18, t19, t28, t30, t37, t38$ can influence the marking of the places contained in the predicate. The resulting possible steps are $\xi_3^p = \{t18\}; \xi_4^p = \{t19\}; \xi_5^p = \{t18, t28\}; \xi_6^p = \{t19, 30\}; \xi_7^p = \{t37\}; \xi_8^p = \{t37, t38\}$. There exists no possible step extension path for any of the steps. The pre-predicate of ξ_3^p includes the source place v_down of the condition arc beside the pre-place of the transition $t18$. The places $ec\_nr, 1B1off, 1B2off$ are not influenced and included in the pre-predicate $ZP'_{\xi_3} = lc\_up \land \overline{1B1off} \land \overline{1B2off} \land \overline{lc\_path} \land v\_down \land$

$ec\_nr \wedge 2B1on$. An extension with event synchronizations reveals a conflict within the resulting predicate $lc\_up \wedge 1B1off \wedge 1B1off \wedge \overline{1B2off} \wedge \overline{lc\_path} \wedge v\_down \wedge ec\_nr \wedge 2B1on$. It is therefore discarded. Two other pre-predicates are discarded because of conflicts:
$ZP'_{\xi^4} = lc\_down \wedge \overline{lc\_path} \wedge 1B1off \wedge 1B2off \wedge \overline{1B2off} \wedge v\_up \wedge ec\_nr \wedge \overline{2B1on}$ and
$ZP'_{\xi^7} = lc\_path \wedge 1B1off \wedge 1B2off \wedge ec\_retra \wedge \overline{ec\_nr} \wedge ev\_on \wedge \overline{2B1on} \wedge \overline{2B1off}$.
The remaining pre-predicates are determined and extended with all applicable invariants to:
$ZP'_{\xi^5} = lc\_up \wedge 1B1on \wedge \overline{1B1off} \wedge \overline{1B2off} \wedge 1B2on \wedge \overline{lc\_path} \wedge \overline{lc\_down} \wedge v\_down \wedge$
$\overline{dneutral} \wedge ec\_nr \wedge \overline{ec\_retra} \wedge \overline{2B1on} \wedge 2B1off$
$ZP'_{\xi^6} = lc\_down \wedge \overline{lc\_path} \wedge \overline{lc\_up} \wedge 1B1off \wedge \overline{1B1on} \wedge 1B2on \wedge \overline{1B2off} \wedge v\_up \wedge$
$\overline{uneutral} \wedge ec\_nr \wedge \overline{ec\_retra} \wedge \overline{2B1on} \wedge 2B1off$,
$ZP'_{\xi^8} = lc\_path \wedge \overline{lc\_down} \wedge \overline{lc\_up} \wedge 1B1off \wedge \overline{1B1on} \wedge 1B2off \wedge \overline{1B2on} \wedge ec\_retra \wedge$
$\overline{ec\_nr} \wedge ec\_path \wedge ec\_extend \wedge ev\_on \wedge \overline{ev\_off} \wedge 2B1on \wedge \overline{2B1off}$.
The analyzed possible steps are uncontrollable, and the next step is the forward step analysis of the pre-predicates ZP'. The enabled transitions of the predicate's places to $lc\_up \wedge 1B1on \wedge \overline{1B1off} \wedge \overline{1B2off} \wedge 1B2on \wedge \overline{lc\_path} \wedge \overline{lc\_down} \wedge v\_down \wedge \overline{dneutral} \wedge ec\_nr \wedge \overline{ec\_retra} \wedge \overline{2B1on} \wedge 2B1off$ are $t18, t28, t38, t15, t36, t39$. All of them are not disabled by other predicate's places and the markings of the pre-places are all given by a state atom. However, just the transition $t15$ is controllable. The application of the forward determination rule leads to the predicate being $lc\_up \wedge 1B1on \wedge \overline{1B1off} \wedge \overline{1B2off} \wedge 1B2on \wedge \overline{lc\_path} \wedge \overline{lc\_down} \wedge d\_neutral \wedge \overline{v\_down} \wedge ec\_nr \wedge \overline{ec\_retra} \wedge \overline{2B1on} \wedge 2B1off$. This follower predicate is not covered by any forbidden state predicate or forcing predicate and cannot constitute a cycle. Therefore, the $FCC = \{(lc\_up \wedge 1B1on \wedge \overline{1B1off} \wedge \overline{1B2off} \wedge 1B2on \wedge \overline{lc\_path} \wedge \overline{lc\_down} \wedge d\_neutral \wedge \overline{v\_down} \wedge ec\_nr \wedge \overline{ec\_retra} \wedge \overline{2B1on} \wedge 2B1off)\}$ is formed.

The complete result of the synthesis is presented in Figure 7.4. The dotted arrows symbolize forward steps while the black ones are backward steps. The gray steps and predicates would result from an extension of the synthesis algorithm described later in this section.

Incomplete observability is considered in the same way as explained in Section 6.3. This step is not further discussed in detail to avoid repetition. From here, explanations continue to use only observable control functions.

The monolithic control functions are assembled of the control components in the next step. The state predicates of all control components referring to the same control input are combined disjunctively to build the control function to the input. Based on the results in Figure 7.4, that leads to the following observable control function to the lifting unit

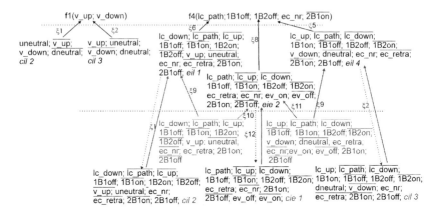

Figure 7.4: Forcing/locking safety controller synthesis results for the example model and the specifications f1 and f4.

input $cil2$:

$$cil2 := 0 \; if \; (uneutral \wedge \overline{v\_up} \wedge v\_down \wedge \overline{dneutral}, \vee$$

$$1B1off \wedge \overline{1B1on} \wedge 1B2on \wedge \overline{1B2off} \wedge uneutral \wedge \overline{v\_up} \wedge \overline{2B1on} \wedge 2B1off) = 1.$$

The monolithic functions are then distributed to local decision functions and commu-nication functions. The monolithic functions contain state atoms to local observable plant parts and nodes of other units (external state atoms). The external atoms are cut out of the control function predicates and substituted by serially numbered commu-nication variables, state predicate by state predicate. Therefore, the affiliation informa-tion are used as saved before composition. The external state atoms referring to the same unit constitute the communication function defining the value of the variables that are substituted. Every communication function results from exactly one predicate of a control function and is a predicate too. That means for the example of $cil2$ that the places $uneutral, v\_up, v\_down, 1B1off, 1B2on, 1B2off, uneutral$ are members of the lift-ing unit as the input $cil2$ is. The first predicate is a local predicate. The second one contains the external state atom $\overline{2B1on}$ to be substituted. The resulting decision function is:

$$cil2 := 0 \; if \; (uneutral \wedge \overline{v\_up} \wedge v\_down \vee$$

$$1B1off \wedge \overline{1B1on} \wedge 1B2on \wedge \overline{1B2off} \wedge uneutral \wedge \overline{v\_up} \wedge com_1) = 1.$$

The generated communication function is $com_1 := \overline{2B1on}$ and is one of the functions of the ejection unit controller. All distributed functions are listed in Table 7.1. The resulting functions are reduced using the p-invariants. The marking of all places in an invariant with the sum 1 is defined completely by the place holding the marking 1. Practically, the reduction also fits to the normal binary communication using both values zero and one as one signal.

Table 7.1: Reduced local observable control functions according to $f1$ and $f4$.

lifting unit		
forcing control functions		
$eil1 = 1$	if	$1B1off \wedge 1B2on \wedge v\_up \wedge com_1 = 1$
$eil4 = 1$	if	$1B1on \wedge 1B2off \wedge v\_down \wedge com_1 = 1$
locking control functions		
$cil2 = 0$	if	$uneutral \wedge v\_down = 1$
$cil3 = 0$	if	$v\_up \wedge dneutral = 1$
communication function		
$com_2 = 1$	if	$1B1off \wedge 1B2off = 1$
ejection unit		
forcing control function		
$eie2 = 1$	if	$com_2 \wedge ev\_on \wedge 2B1on = 1$
communication function		
$com_1 = 1$	if	$2B1off = 1$

The given algorithm for forcing/locking controller synthesis leads to control functions alternatively locking or forcing the system. From the formal point of view, that is a correct solution. But the plant element changing its state by forcing can get switched back by other control components. This is mostly unwanted for real systems. It seems to be a necessary extension of the algorithm to disable the steps from a follower predicate of a forcing state back to this predecessor state.

There are two possible procedures to include that extension. First, the backward search could be proceeded for forcing states $(ZP \in FCC)$ until a pre-predicate covering the follower state $(ZP^*with(ZP, e^{in}, ZP^*) \in FCC)$ of the forcing state is found. Second, a forward analysis can be performed from the follower marking to the forcing marking, and a locking input can be determined. There should be a controllable trajectory between the follower predicate and the forcing predicate because an actuator with controllability in just one direction makes not much sense. However, the analysis would determine also if such trajectory was missing because the net is bounded. For actuators with two state models

Table 7.2: Extended local observable control functions according to $f1$ and $f4$.

lifting controller		
$eil1 = 1$	if	$1Bloff \land 1B2on \land v\_up \land com_1 = 1$
$eil4 = 1$	if	$1Blon \land 1B2off \land v\_down \land com_1 = 1$
$cil2 = 0$	if	$uneutral \land v\_down \lor 1Bloff \land 1B2on \land uneutral \land com_1 = 1$
$cil3 = 0$	if	$v\_up \land dneutral \lor 1Blon \land 1B2off \land dneutral \land com_1 = 1$
$com_2 = 1$	if	$1Bloff \land 1B2off = 1$
ejection controller		
$cie2 = 0$	if	$com_2 \land ev\_off \land 2Bloff = 1$
$eie1 = 1$	if	$com_2 \land ev\_on \land 2Blon = 1$
$com_1 = 1$	if	$2Bloff = 1$

(as in example in Figure A.3), both possible procedures lead to the same result. The additionally obtained predicates and locking control inputs are displayed in gray in Figure 7.4. They are calculated by backward search for the forcing states. Control predicates are all pre-predicates covering the forcing predicate. The two completely gray predicates in Figure 7.4 are additionally determined but not included in any control function. The results of the algorithm with re-activation prevention (gray function parts) are shown in Table 7.2.

The control functions are distributed to the local controllers of the two units and just two communication functions are necessary for exchange of information in this example.

7.3 Comparison

The presented forcing/locking controller synthesis generates a reduced set of forbidden state predicates in accordance to the pure locking controller synthesis in [MH06b] and Chapter 6. The set of forbidden state predicates in the given example contains the two specification states only, while the result for a pure locking controller leads to five forbidden state predicates. The forcing states are not forbidden and not avoided therefore. Note: Every state predicate characterizes a number of system's states. The reduced number of forbidden state predicates decreases not only the computational complexity, moreover this means that the result is less restrictive to system behavior in converse. The monolithic backward search is maximally permissive but in general the synthesized forcing/locking controllers are not maximally permissive. That is mainly because the reachability of follower predicates to forced steps can not be finally decided without determination of all reachable states. The condition of full description of pre- (post-) region of all step transitions by the forcing state predicate is introduced therefore. In the case that there is a transition set satisfying all other conditions but this, forcing the step could be possible if the follower marking was reachable. So, the forcing step could be added in this case and could result in a less restrictive controller. Further, the permissiveness of the trajectories started by a forced step is not checked and compared to alternative forceable steps being related to the same predicate.

Comparing the locking controller results of the modular backward search (Section 6.2) and the monolithic backward search (7.1.2) ends up with two kinds of efficiency. The modular algorithm features an improved computational efficiency and leads directly to some kind of reduced distributed control functions as described in Section 6.4. The monolithic variant described in Section 7.1.2 uses a composed model and does not take advantage of modular search execution. But it still takes advantage of the use of partial marking, which subsumes a number of markings. Moreover, the principle of backward search within both approaches avoids calculation of state transitions not influencing the marking of specified places. That means, only trajectories leading to forbidden marking are calculated. Further, the monolithic backward search as described in this chapter is maximally permissive, which is the most efficient result. That means for implementations, the safety functions are minimal restrictive and leave a maximal feasible plant behavior for constitution of the desired processes by process control functions. The synthesized control functions are not minimal in terms of size. They can contain predicates not covering any reachable state, because reachability is not checked to clamp computational complexity.

An approach for backward search using minimal covering sets is presented in [DM07] for a modification of $_SNCES$ that are called R_SNCES. The use of minimal covering sets would reduce the control predicates but does still not consider the reachability of follower markings to control predicates. Hence, the result would not be minimal in general.

The modular backward search does not consider the forward effects, which results in generally not maximally permissive control functions. The improvement of the modular algorithm to consider forward effects without excessively increased computational complexity must be left for further research. Well structured models as described in Section 4.2 however can contain forward effects only in the workpiece property model. For models without forward effect both approaches result in controller models with the same permissiveness.

The next chapter describes the implementation of synthesized distributed control functions in terms of function blocks according to the international standard IEC 64199.

8 Embedding Control Functions in IEC 61499 Function Blocks

The last step of the model-driven generation of correct controller code is the implementation of the controller model in an executable software code (see Figure 1.2). A control code is executable as it is written in a language directly executable on industrial controller hardware or importable to platform-depending engineering tools able to compile code executable on the hardware. The last is common in programming industrial controllers. Two different international standards define the design of industrial controller software and programming languages. First, the international standard [IEC61131-3] defines a set of programming languages for Programmable Logic Controllers (abbr. PLC). The early-aged standard [IEC61499] combines some of the IEC 61131 languages with a new event-driven execution model. It is strictly component-based and allows the integration high level programming languages as JAVA or C/ C++.

The international standard IEC 61499 provides concepts for design and implementation of distributed control systems on distributed control hardware configurations. Introductions to the concept can be found in [Vya06, VH99, Chr00]. The standard was developed particularly focusing on distributed controller applications. A major attribute is the strong encapsulation of functionality into Function Blocks (abbr. FB). The primitive of Function Blocks are the so-called Basic Function Blocks (abbr. BFB). Further the standard allows modular and hierarchical composition of system's structure and behavior. The hierarchical composition is realized by Composite Function Blocks (abbr. CFB). The concept formulated in the standard supports the design of applications spreading across different recourses and/or devices. Interaction between FBs is established via signal interconnection of two basic signal types. The event signals are interpreted similar to events in automata or $_sNCES$ and data signals carrying data in a wide range of data types. Applications are built of those elements as a network of interconnected FBs within a graphical development environment. The platform-independent applications can be distributed across different devices and resources. Functional and communication design are realized within the same framework and within the same design step.

This work focuses on the synthesis of correct controller models. The execution of IEC61499 compliant control code depends on the used runtime environment [SZC+06a]. The correct implementation of the code generation according to a given runtime environment would have to be verified. That, however, is out of scope of this work.

The following sections focus on the implementation of safety controller models in Basic Function Blocks and their integration in control applications. The distributed controller model consists of two control function types. Hence, two BFBs are generated for every local controller. The synthesis of safety controllers is just one aspect of controller design. A process controller is supposed to exist to complete the controller structure as introduced in Chapter 3.

First, the basics of BFBs are introduced in the next section, followed by the description of the generation of the BFB elements with exemplification of every step. This chapter is closed by a short summary of the results.

8.1 IEC 61499 Basic Function Blocks

The basic building blocks of each control application are the Basic Function Blocks. They are functional units that encapsulate basic functionalities in terms of algorithms for data processing together with an internal execution control, internal variables and a communication interface as shown in Figure 8.1.

The *interface* of BFBs consists of data in-/outputs and event in-/outputs. The values of data inputs of different types together with internal variables are processed by algorithms. Data outputs provide information of currently processed data to the environment. *Internal variables* are not accessible outside the FB. Occurring events at the event inputs trigger the execution of *algorithms* by triggering state changes within the *Execution Control Chart* (abbr. ECC). Event outputs propagate the termination of the algorithms and the availability of newly processed data to the environment. Additionally, the interface contains *associations of event and data inputs* as well as of event and data outputs. The associations relate an incoming/outgoing event to the update of associated the input data value and the value of data outputs, respectively. That means, an input data value is read from the data source only when an event appears, which is associated to the input. Data outputs are updated to the internally computed values with associated outgoing events only.

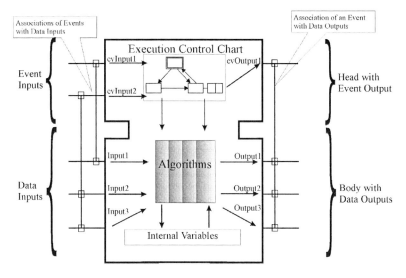

Figure 8.1: Structure of Basic Function Blocks.

The execution control inside each BFB is specified by an Execution Control Chart. The ECC of a BFB is a state machine with state transitions and states with actions. The transitions are labeled by conditions, which can be an event, a Boolean function of guard variables or the Boolean combination of both. A guard variable is an input, output or internal variable of Boolean data type. The actions associated to a state can be algorithms and events induced after termination of the algorithm's execution. Every BFB contains exactly one ECC.

Algorithms process input and internal data and assign output data. Function block diagrams, ladder logic, structured text and higher programming languages, depending on the executing runtime environment, can be used to formulate algorithms. Only structured text algorithm are used for the control function implementation in this contribution.

For sake of completeness may be noted that *Composite Function Blocks* (abbr. CFB) consist of a network of FBs and an interface identical to that of BFBs.

Table 8.1: Example set of local control functions (example in Section 6.3).

ejection unit		
$cie1 = 0$	if	$2B1on \wedge ev\_off \wedge (com_1^+ \vee com_2^+)$
$com_1^- = 1$	if	$\overline{2B1on} \wedge com_2^+ \vee 2B1on \wedge ev\_on \wedge com_2^+$
$com_2^- = 1$	if	$\overline{2B1on} \wedge com_1^+ \vee 2B1on \wedge ev\_on \wedge com_1^+$

8.2 Transformation

The synthesized safety controller model consists of decision functions and communication functions. One Decision Control FB (*abbr. DCFB*) and one Communication FB (abbr. CmFB) has to be generated for every local controller. The DCFB implements the local safety control functions as constituted in the structure shown in Figure 3.1. The *DCFB* is grouped as a filter between a separately designed or existing local process control Function Block and the controller outputs as depicted in Figure 8.2 (*ejection decision FB*). It locks the controller outputs to the current value as long as the assigned safety control function is satisfied. The value of the output of the process control FBs is passed in any other case.

The *CmFB* implements the communication functions and defines the values of communication variables depending on the local observation and the values of communication variables provided by other local controllers.

In the following, the generation is described for every FB element for DCFB and CmFB separately. Only internal variables are not discussed and generated because all functions are state-based and decided by the inputs values. Hence, no memory is necessary. The generation is exemplified step by step using the decision and communication functions for the ejection unit given in Section 6.3 and listed in Table 8.1. The generation of signal interconnections within the system configuration is not addressed in this work. The whole design is based on an important but simple design rule:

Every updated value of a data output of a FB connected to one of the local safety controller FBs is confirmed by an event labeled with CNF.

The same also holds for the safety controller FBs if generated as described in the following.

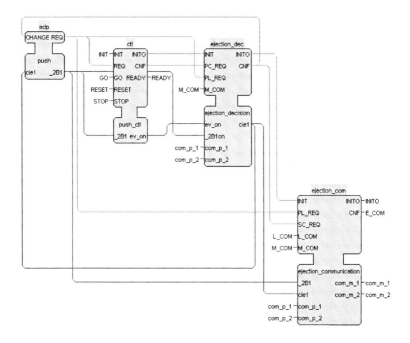

Figure 8.2: Structure of the local ejection controller with generated safety control function blocks.

8.2.1 Interface

The event and data inputs and outputs as well as their associations are generated to define the interface of the FBs. All inputs and outputs of the generated FBs are taken over from the synthesized control functions, except the generally used initialization event input (*INIT*) and output (*INITO*).

DCFBs have one additional event output, the *CNF*. It confirms new values at the data outputs. One data output is generated for every *controller output*. The set of controller outputs has to contain the modeled set of plant inputs. Not all of the outputs are safety-controlled, but all are connected trough the DCFB to admit a consistent control structure.

An DCFB has always the three event inputs *INIT*, *PC_REQ* and *PL_REQ*. The denotation PC_REQ stands for process control request. This input is activated if a process controller FB confirms its outputs. The event input PL_REQ (plant request) is activated if a controller input for plant observation is refreshed. Additionally, one communication request event input is generated for every local controller sending communication variables contained in a local decision function. The construction of the DCFBs interface considers the following rules.

A data input is generated for:

- all local controller inputs (local plant observation) being part of a local decision function associated with the events $INIT$ and PL_REQ,

- all process controller outputs (connected with the associated process controller data output) associated with the events $INIT$ and PC_REQ and

- all communication variables being part of a local decision function associated with the communication event $*\_COM$ of the providing unit controller.

A data output is generated for:

- all local controller outputs.

All data inputs and outputs are of Boolean data type.

Example:

The example decision function $cei1$ contains communication variables provided by the measuring unit CmFB. The event input M_COM therefore is created beside INIT, PC_REQ and PL_REQ. Further, the data inputs $2B1on$, ev_off, com_1^+, com_2^+ are created. The interface of the $ejection\_dec$ FB with the associations is shown in Figure 8.4(a) for the example decision function.

The only controller output $cie1$ of the ejection unit control is safety-controlled and is the only data output from the ejection process control FB in the example.

The inputs to be generated for $CmFBs$ differ for different communication functions to be implemented. The two different synthesis methods result in different communication structures. The CmFBs have just plant observations and safety controller output values as inputs in case the functions result from the homogeneous approach (see Chapter 7). The communication functions additionally contain communication variables if synthesized with the modular synthesis (see Chapter 6). The $CmFBs$ just have the event inputs $INIT$ and PL_REQ, SC_REQ (confirming new safety controller data output values) in the first case. The FB only has plant signal data and DCFB output data inputs in that case. If the communication functions contain communication variables provided by

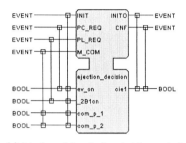
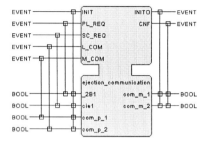

(a) Interface of the ejection decision control FB for the example function in Table 8.1.

(b) Interface of the ejection communication FB for the example function in Table 8.1.

Figure 8.3: Interfaces of the example FBs.

other local controllers, then the inputs for the communication request events for all sending controllers, the communication variable data inputs and their associations are generated as described for DCFBs additionally to the mandatory $INIT$, PL_REQ and SC_REQ events. The event outputs of an CmFB are just like the event outputs of DCFBs. The CNF event represents the communication confirmation for controllers receiving communication variables.

A data output for the CmFB is generated for:

- all communication variables assigned by a local communication function.

The example communication functions are of the second type and thus one communication event input and data inputs are generated. The resulting CmFB for the ejection controller is displayed in Figure 8.4(b).

Next, the transformation of control functions to algorithms is described.

8.2.2 Algorithms

The algorithms are the main element of implementation of the controller model. They have to assign the local safety controller FB outputs as modeled. The algorithms are formulated using the structured text instruction IF.

First, the algorithms in $DCFBs$ are described. One algorithm is defined for every local controller output. Algorithms to outputs not influenced by a safety controller function simply assign the input value from the process control FB to the output of the DCFB.

Safety controller-influenced outputs depend on the process control FBs output values and the synthesized decision functions.

The functions are implemented by an instruction of the general structure:

$IF\,![\text{decision function}]\,then\,[\text{process control output}];$

Naturally, the decision function is the condition of the instruction. If the condition is satisfied, then a safety control action is required, and the safety controller output value is locked until it is not satisfied anymore. If it is not satisfied, then the value is assigned to the value of the input named as the controlled output. So, the value of the process controller output is assigned to the local controllers output. Example:

Hence, the example algorithm for the $cei1$ output is generated as follows.

$$IF\ (NOT(NOTev\_on\ \&\ \_2B1on\ \&\ (com\_p\_1\mid com\_p\_2)=true))$$
$$\text{then } cie1 := ev\_on;\ \text{END\_IF};$$

The algorithms for $CmFBs$ differ just in the assigned values. Again, one algorithm is defined for every local communication function. The communication function is used as condition. The value of the communication variable (named as defined in the controller model) is assigned to $true$ if the condition is true. If not, the value of the variable is $false$. The resulting structure is:

$$IF[\text{communication func.}]\,then[\text{communication var.}] := true;$$
$$ELSE[\text{communication var.}] := false;$$

Example:

The two generated algorithms for the example in Table 8.1 are:

- com_m_1

  $$IF\ (NOT\_2B1\ \&\ com\_p\_2\mid\ \_2B1\ \&\ cie1\ \&\ com\_p\_2=true)\ \text{then}$$
  $$com\_m\_1 := true;$$
  $$\text{ELSE } com\_m\_1 := false;\ \text{END\_IF};$$

- com_m_2

  $$IF\ (NOT\_2B1\ \&\ com\_p\_1\mid\ \_2B1\ \&\ cie1\ \&\ com\_p\_1=true)\ \text{then}$$
  $$com\_m\_2 := true;$$
  $$\text{ELSE } com\_m\_2 := true;\ \text{END\_IF};$$

The execution of the algorithms is started depending on the ECC's state. The rules to generate the ECC are introduced next.

8.2.3 Execution Control Chart

The execution of the algorithms is organized by the ECC. The trivial way would be to execute all algorithms when one of the incoming events occurs. Instead, a more proper design is defined to achieve more efficient execution in most cases. The procedure for ECC construction is identical for CmFBs and DCFBs. The idea is to execute only algorithms containing variables associated to the occurred event. That complies with the idea of event-driven execution in the IEC 61499.

Every ECC is generated with three default states and a number of states depending on the control functions to be implemented. An initial *WAIT* state, an *INIT* state and the *PC_REQ* state are defined and as many additional states as the FB has event inputs, while the identifier of the state equals the event's name.

No algorithms are executed in the initial *WAIT* state. All algorithms have to be executed in the mandatory initialization state *INIT* and the *PC_REQ* state. Thus, the safety specifications are checked within initialization and when the process controller requests to change the value of a controller output. Last is necessary, because actuator information and therefore controller output values are components of the decision function (see Section 4.3).

The last associated algorithm of every state is assigned with the *CNF* output event. That holds for all generated states except for *WAIT*. The event is associated with the last algorithm to ensure the completion of all algorithms before confirming the execution, independently of the used runtime environment. The processing of events and firing of state-associated events differs in various runtime implementations and execution models according to the IEC 61499 as described in [DV06, SZC+06b]. An always enabled transition is generated from any created algorithm-executing state to *WAIT*. There are no state sequences in the implementation of the state-based controller model.

The following construction rule holds for all states excluding *WAIT* and *INIT*. The generation of states and state transitions differs in two cases according to the number of algorithms in the FB.

1. If there is only one algorithm to be executed in the FB, then just the *PC_REQ* state is generated. One transition is generated for every event input associated to an input

data variable contained in the function implemented by the algorithm, except for the
INIT event. Every transition is assigned with one of the events as condition.

2. If there are more than one algorithms to be contained in the FB, i.e. more than one
   function to be implemented, then one additional state (execution state) is generated
   for every event input beside the mandatory state INIT. One transition from *WAIT*
   to that created state is generated. It is assigned with the event as condition. The
   state is named as the initiating event. Algorithms containing any variable associated
   to the event are added to the actions executed in the state. That means for example, algorithms including one or more local observations are assigned to the state
   PL_REQ.

Example:
The construction of the ECC is exemplified with the ejection unit example again. There
is just one decision algorithm synthesized. Hence, three transitions to the execution state
and one back to the *WAIT* state are generated as shown in Figure 8.4(a).

The ECC of *local communication FBs* is generated according to the same rules. More than
one communication function has to be implemented in the example of the ejection unit.
Four execution states are generated for the four event inputs beside *INIT*, according to the
second case listed on page 119. The resulting ECC is shown in Figure 8.4(b).

That structure ensures the execution of only those algorithms, which include functions
influenced by the change of a value noticed by an event.

Obviously, the ECC can be reduced by determination of states executing the same actions.
That would require an adopted rule for transition generation for concentrated states.

Beside the generation of the ECC nodes, arcs and their relation, also *graphical information*
has to be generated. An example for format guidelines as used in the example in Figure 8.5
is given in the following. A placement of the states in three rows is chosen because of the
given, always identical, basic structure of the generated ECCs. The used alignment and
dimensions are displayed in Figure 8.5. The center point is the *WAIT* state. The INIT
state is located directly above and the execution states in a row below. The example
distances in Figure 8.5 are those chosen for the example of the ejection unit (ECCs in
Figure 8.4, fbt-description in Listing A.3.2).

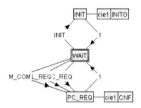

(a) ECC of the ejection decision control FB for
the example function in Table 8.1.

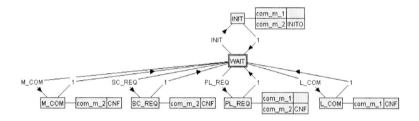

(b) ECC of the ejection communication FB for the example function in Table 8.1.

Figure 8.4: ECCs of the example FBs.

8.3 Use and extension

The transformation is described for locking control function in this chapter. The decision
Function Blocks act as a guard between process control or manual operator actions and
the controller output to the plant. It checks if an output change is permitted under the
given situation. The implementation of forcing safety control functions induce to generate
an additional Function Block. It reads plant outputs only and influences the controller
outputs in active manner. It will change the controller output values if required under the
given plant observation. This forcing safety control acts kind of parallel to the process
control FBs. Both safety control functions reduce the behavior attainable by the process
control action. This has to be considered in the process control function design.

The generated algorithms and the whole FBs are of relatively simple structure and readable

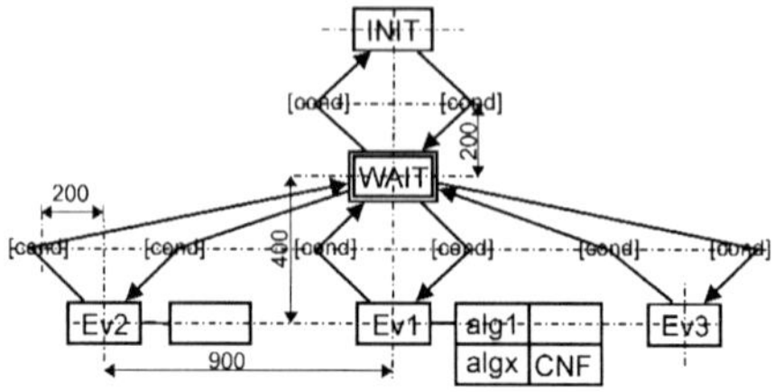

Figure 8.5: Example for the general alignment of nodes and arcs of a generated ECC.

by developers familiar with IEC 61499. The use of FB types allows fast update of an existing control application by adding the generated safety FBs. It is only necessary to perform a new automatic synthesis of the FBs in case of changed safety specifications or model properties. If the interfaces remain unchanged, then no change is needed in the control application. The proposed integration of the generated FBs into a control structure allows the use of safety function controller synthesis independently of the concrete implementation of the process controller components. It can therefore also be inserted as extension of existing controller applications.

9 Conclusions and Outlook

Control synthesis and especially distributed control synthesis are a major challenge of scientific research and practical application. The reduction of computational complexity turns out to be the main requirement for real-scale applications. A systematic and efficient modeling technique is a prerequisite for the application of formal synthesis. The use of a modular and hierarchical model supports this design requirements as well as the synthesis of distributed control functions.

This work shows that using partial markings and backward search offers major advantages. It reduces the complexity in terms of numbers of analysis steps by subsuming a number of states in one predicate. The backward search focuses the analysis only on net paths really influencing markings subsumed by the analyzed predicates. The reduced complexity is shown by a comparison of the example results with the size of the reachability graph. Additional reduction of the complexity can be gained from the modular synthesis. The modular approach further leads to distributed control functions directly. The comparison with the homogeneous approach using backward search also shows a reduced number of computed predicates. Both approaches avoid the enumeration of the complete reachability set.

Proofs of correctness were given for the used possible step definitions and the introduced monolithic synthesis approach. The modular approach is not maximal permissive in general, but the maximal permissiveness of the monolithic backward search is proven.

The proposed methods show potentials that may bring synthesis a significant step towards feasibility to systems of realistic size and complexity.
The potential benefits are significant:

1. The algorithms have less computational complexity than algorithms determining control functions based on the complete reachable state space. Additional reduction is achieved by the modular algorithm. The gain one could get obviously depends on the structure of the plant model and the assignment of controllers to it (e.g. grade of distribution). This, in turn, depends on the decision of human beings.

2. The plant model is as close to reality as needed to be the origin of correct results. It can be constructed systematically from predefined modules rather than designed from scratch. This helps human and improves reliability and performance during the modeling and validation process.

3. The presented modular synthesis approach can be run partially in parallel. This aspect is remarkable and may become more interesting for further application to systems of realistic scale.

4. The combination of forcing and locking control action reduces improves the permissiveness of safety controller according to the same specification.

In contrast to earlier works, the described approaches consider incomplete controllability as well as incomplete observability. This is indispensable for application on real control problems. Also the generation of executable controller code in terms of IEC 61499 compliant Basic Control Function Block is discussed. The given example shows, how Function Blocks support a straightforward implementation of the distributed controller model functions. The placement of the safety Function Blocks as a kind of filter protects against programming faults in the process control functions and against operating error.

This work describes a solution for all steps of the synthesis sequence stated in Figure 1.2. Nonetheless, there are still open issues on the way to apply controller synthesis for real applications. The extension on specifications of forbidden state or state transition sequences would allow to cover a wider range of control problems.

A major step to fully automated controller code synthesis would be achieved by synthesis approaches for process control functions. This aspect is not touched by this work. However, the reviewed model definitions and extended methods for structural model analysis also improve the basis for future work on this topic.

Another field of computer science should be involved, when it comes to implementation of controller synthesis and code generation. The implementations have to be verified on the formal described behavior to take the full advantage of the use of formal methods. Only the use of verified software would finally guarantee correctness of the synthesis results. Correctness in this context always refers to the given model and specification.

If finally all formal and technical issues were solved, the final point left would be to convince involved engineers of the trustability of the method and the advantage to gain from the additional task of plant modeling. This statement is based on a number of discussions the author had with control experts. The assignment of safety controller function synthesis to Safety Instrumentation Systems has high requirements but would give a starting point for the introduction of a synthesis method to real application. The design effort currently practiced on SIS applications is very high. The use of controller synthesis could reduce the cost and improve the quality of implementations when the mentioned open issues get solved.

A Appendix

A.1 Tables

Table A.1: Controllable and observable nodes of the lifting, ejection and measuring unit.

lifting unit	ejection unit	measuring unit
controllable transition		
$t13, t14, t15, t16$	$t32, t33$	$t42, t43$
controlled places (observable)		
$v\_up, uneutral, dneutral,$ v_down	ev_on, ev_off	mv_on, mv_off
observable places		
$1B1on, 1B1off, 1B2on, 1B2off$	$2B1on, 2B1off$	$3B1on, 3B1off, hs\_on,$ hs_off

Table A.2: Modular synthesized local control functions to the specified forbidden states $f3$
and $f4$ (example in Section 6.3).

lifting unit		
$cil2 = 0$	if	$1B1off \wedge 1B2on \wedge uneutral \wedge com_2^-$
$cil3 = 0$	if	$1B1on \wedge 1B2off \wedge dneutral \wedge com_2^-$
$com_2^+ = 1$	if	$1B1off \wedge 1B2off \ \vee \ 1B1off \wedge 1B2on \wedge v\_up \ \vee \ 1B1on \wedge$ $1B2off \wedge v\_down$
ejection unit		
$cie1 = 0$	if	$2B1on \wedge ev\_off \wedge (com_1^+ \vee com_2^+)$
$com_1^- = 1$	if	$\overline{2B1on} \wedge com_2^+ \ \vee \ 2B1on \wedge ev\_on \wedge com_2^+$
$com_2^- = 1$	if	$\overline{2B1on} \wedge com_1^+ \ \vee \ 2B1on \wedge ev\_on \wedge com_1^+$
measuring unit unit		
$cim1 = 0$	if	$3B1off \wedge mv\_off \wedge com_1^-$
$com_1^+ = 1$	if	$\overline{3B1off} \ \vee \ 3B1off \wedge mv\_on$

A.2 Figures

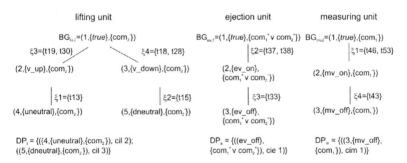

Figure A.1: Results of the modular backward search without extension by event synchronizations based on the forbidden states $f3$ and $f4$ (as in the example Section
6.3).

Figure A.2: Results of the monolithic backward search with extension by event synchronizations based on the forbidden states $f3$ and $f4$.

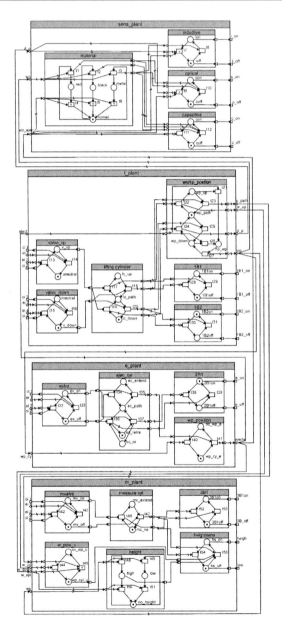

Figure A.3: $_sNCES$ model of the testing station.

A.3 Function Block examples

A.3.1 Ejection decision

The following listing shows the *.fbt* file implementing the decision function of the example used in Section 6.3.

```xml
<?xml version="1.0" encoding="UTF-8"?>
<!DOCTYPE FBType SYSTEM "http://www.holobloc.com/xml/LibraryElement.dtd" >
<FBType Name="ejection_decision" Comment="" >
  <Identification Standard="61499-2" />
  <VersionInfo Organization="Martin-Luther-University" Version="0.1" Author="DM" Date="
    2010-02-02" Remarks="" />
  <InterfaceList>
    <EventInputs>
      <Event Name="INIT" Comment="Initialization_Request" >
        <With Var="ev_on" />
        <With Var="_2Blon" />
        <With Var="com_p_1" />
        <With Var="com_p_2" />
      </Event>
      <Event Name="PC_REQ" Comment="Normal_Execution_Request" >
        <With Var="ev_on" />
      </Event>
      <Event Name="PL_REQ" Comment="Request_from_plant" >
        <With Var="_2Blon" />
      </Event>
      <Event Name="M_COM" >
        <With Var="com_p_1" />
        <With Var="com_p_2" />
      </Event>
    </EventInputs>
    <EventOutputs>
      <Event Name="INITO" Comment="Initialization_Confirm" >
        <With Var="cie1" />
      </Event>
      <Event Name="CNF" Comment="Execution_Confirmation" >
        <With Var="cie1" />
      </Event>
    </EventOutputs>
    <InputVars>
      <VarDeclaration Name="ev_on" Type="BOOL" Comment="plant_obs" />
      <VarDeclaration Name="_2Blon" Type="BOOL" Comment="plant_obs" />
      <VarDeclaration Name="com_p_1" Type="BOOL" Comment="communication_variable" />
      <VarDeclaration Name="com_p_2" Type="BOOL" Comment="communication_variable" />
    </InputVars>
    <OutputVars>
      <VarDeclaration Name="cie1" Type="BOOL" Comment="controlled_output" />
    </OutputVars>
  </InterfaceList>
  <BasicFB>
    <ECC >
```

```
<ECState Name="WAIT" Comment="Initial_State" x="2800" y="800" >
</ECState>
<ECState Name="INIT" Comment="Initialization" x="2800" y="400" >
  <ECAction Algorithm="cie1" Output="INITO" />
</ECState>
<ECState Name="PC_REQ" Comment="Normal_execution" x="2800" y="1200.0" >
  <ECAction Algorithm="cie1" Output="CNF" />
</ECState>
<ECTransition Source="WAIT" Destination="INIT" Condition="INIT" x="2600" y="600" />
<ECTransition Source="INIT" Destination="WAIT" Condition="1" x="3000" y="600" />
<ECTransition Source="WAIT" Destination="PC_REQ" Condition="PC_REQ" x="2600" y="1000
    " />
<ECTransition Source="WAIT" Destination="PC_REQ" Condition="PL_REQ" x="2400" y="1000
    " />
<ECTransition Source="WAIT" Destination="PC_REQ" Condition="M_COM" x="2200" y="1000"
    />
<ECTransition Source="PC_REQ" Destination="WAIT" Condition="1" x="3000" y="1000" />
</ECC>
<Algorithm Name="INIT" Comment="Initialization_algorithm" >
<ST Text="" />
</Algorithm>
<Algorithm Name="cie1" Comment="Normally_executed_algorithm" >
<ST Text="IF_(NOT(NOT_cie1_&#38;_2B1on_&#38;_(com_p_1_|_com_p_2)_=_true))_then&#10;
    cie1_:=_false;&#10;ELSE&#10;cie1_:=_cie1;&#10;END_IF;&#10;" />
</Algorithm>
</BasicFB>
</FBType>
```

A.3.2 Ejection communication

The following listing shows the *.fbt* file implementing the communication functions of the
example used in Section 6.3.

```
<?xml version="1.0" encoding="UTF-8"?>
<!DOCTYPE FBType SYSTEM "http://www.holobloc.com/xml/LibraryElement.dtd" >
<FBType Name="ejection_communication" Comment="communication_of_the_safety_controller" >
  <Identification Standard="61499-2" />
  <VersionInfo Organization="Martin-Luther-University" Version="0.1" Author="DM" Date="
    2010-02-02" Remarks="" />
  <InterfaceList>
    <EventInputs>
      <Event Name="INIT" Comment="Initialization_Request" >
        <With Var="_2B1" />
        <With Var="cie1" />
        <With Var="com_p_1" />
        <With Var="com_p_2" />
      </Event>
      <Event Name="PL_REQ" Comment="Plant_Request" >
        <With Var="_2B1" />
      </Event>
      <Event Name="SC_REQ" Comment="Control_Request" >
```

```xml
        <With Var="ciel" />
      </Event>
      <Event Name="L_COM" >
        <With Var="com_p_2" />
      </Event>
      <Event Name="M_COM" >
        <With Var="com_p_1" />
      </Event>
    </EventInputs>
    <EventOutputs>
      <Event Name="INITO" Comment="Initialization_Confirm" >
        <With Var="com_m_1" />
        <With Var="com_m_2" />
      </Event>
      <Event Name="CNF" Comment="Execution_Confirmation" >
        <With Var="com_m_1" />
        <With Var="com_m_2" />
      </Event>
    </EventOutputs>
    <InputVars>
      <VarDeclaration Name="_2B1" Type="BOOL" Comment="Sensor_Input_3B1" />
      <VarDeclaration Name="ciel" Type="BOOL" />
      <VarDeclaration Name="com_p_1" Type="BOOL" Comment="com_variable_from_measuring_unit
          " />
      <VarDeclaration Name="com_p_2" Type="BOOL" Comment="com_variable_from_lifting_unit"
          />
    </InputVars>
    <OutputVars>
      <VarDeclaration Name="com_m_1" Type="BOOL" Comment="Output_event_qualifier" />
      <VarDeclaration Name="com_m_2" Type="BOOL" />
    </OutputVars>
  </InterfaceList>
  <BasicFB>
    <ECC >
      <ECState Name="WAIT" Comment="Initial_State" x="2800" y="800" >
      </ECState>
      <ECState Name="INIT" Comment="Initialization" x="2800" y="400" >
        <ECAction Algorithm="com_m_1" />
        <ECAction Algorithm="com_m_2" Output="INITO" />
      </ECState>
      <ECState Name="PL_REQ" Comment="Normal_execution" x="2800" y="1200" >
        <ECAction Algorithm="com_m_1" />
        <ECAction Algorithm="com_m_2" Output="CNF" />
      </ECState>
      <ECState Name="SC_REQ" x="1900" y="1200" >
        <ECAction Algorithm="com_m_2" Output="CNF" />
      </ECState>
      <ECState Name="L_COM" x="3700" y="1200" >
        <ECAction Algorithm="com_m_1" Output="CNF" />
      </ECState>
      <ECState Name="M_COM" x="1000" y="1200" >
        <ECAction Algorithm="com_m_2" Output="CNF" />
      </ECState>
      <ECTransition Source="WAIT" Destination="INIT" Condition="INIT" x="2600" y="600" />
```

```xml
    <ECTransition Source="INIT" Destination="WAIT" Condition="1" x="3000" y="600" />
    <ECTransition Source="WAIT" Destination="PL_REQ" Condition="PL_REQ" x="2600" y="1000
        " />
    <ECTransition Source="PL_REQ" Destination="WAIT" Condition="1" x="3000" y="1000" />
    <ECTransition Source="WAIT" Destination="SC_REQ" Condition="SC_REQ" x="1700" y="1000
        " />
    <ECTransition Source="SC_REQ" Destination="WAIT" Condition="1" x="2100" y="1000" />
    <ECTransition Source="WAIT" Destination="L_COM" Condition="L_COM" x="3500" y="1000"
        />
    <ECTransition Source="L_COM" Destination="WAIT" Condition="1" x="3900" y="1000" />
    <ECTransition Source="WAIT" Destination="M_COM" Condition="M_COM" x="800" y="1000" /
        >
    <ECTransition Source="M_COM" Destination="WAIT" Condition="1" x="1200" y="1000" />
  </ECC>
 <Algorithm Name="com_m_1" >
  <ST Text="IF_(NOT__2B1_&#38;_com_p_2_|__2B1_&#38;_cie1_&#38;_com_p_2_=_true)_then&#10;
    com_m_1_:=_true;&#10;ELSE&#10;com_m_1_:=_false;&#10;END_IF;&#10;" />
 </Algorithm>
 <Algorithm Name="com_m_2" Comment="Normally_executed_algorithm" >
  <ST Text="IF_(NOT__2B1_&#38;_com_p_1_|__2B1_&#38;_cie1_&#38;_com_p_1_=_true)_then&#10;
    com_m_2_:=_true;&#10;ELSE&#10;com_m_2_:=_true;&#10;END_IF;&#10;" />
 </Algorithm>
 </BasicFB>
</FBType>
```

Bibliography

Normative References

[IEC61131-3] IEC 61131-3:2003, Programmable controllers - Part 3: Programming languages, 2003.

[IEC61499] IEC 61499:2005, Function blocks, 2005.

[IEC61508-0] IEC 61508-0:2005, Functional safety of E/E/PE safety-related systems - Part 0: Functional safety and IEC 61508, 2005.

[IEC61511] IEC61511:2005, Functional safety - Safety instrumented systems for the process industry sector, 2005.

[IEC62061] IEC62061:2005, Safety of machinery - Functional safety of safety-related electrical, electronic and programmable electronic control systems, 2005.

[ISO/IEC199775] ISO/IEC 199775, Information technology - Computer graphics and image processing - Extensible 3D (X3D), 2004.

[OMGSysML1.1] OMGSysML 1.1: OMG Systems Modeling Language, Version 1.1, 2008.

Online-References

[TNCES] TNCES Workbench. *http://sourceforge.net/projects/tnces-workbench/*.

[example] Example Testing Station. *http://aut.informatik.uni-halle.de/forschung/synthese/example/*.

[testbed] Modular testbed for distributed control. *http://aut.informatik.uni-halle.de/forschung/testbed/index.de.php*.

Literature

[BK90]   Z. Banaszak and B. H Krogh. Deadlock avoidance in flexible manufacturing
          systems with concurrently competing process flows. *IEEE Transactions on
          Robotics and Automation*, 6(6):724–734, 1990.

[BL98]   G. Barrett and S. Lafortune. On the synthesis of communicating controllers
          with decentralized information structures for discrete-event systems. In *Pro-
          ceedings of the 37th IEEE Conference on Decision & Control, Tampa, Florida,
          USA*. IEEE, December 1998.

[Bra94]   A. Brandstaedt. *Graphen und Algorithmen*. Teubner Verlag, Stuttgart, 1994.

[Bre05]   R. W. Brennan. *An Initial Automation Object Repository for OOONEIDA*,
          volume 3593/2005, chapter Holonic and Multi-Agent Systems for Manufactur-
          ing, pages 154–164. Springer Berlin / Heidelberg, 2005.

[CDFV88] R. Cieslak, C. Desclaux, A.S. Fawaz, and P. Varaiya. Supervisory control of
          discrete-event processes with partial observation. In *IEEE Transactions on
          Automatic Control*. IEEE, March 1988.

[Chr00]   J.H. Christensen. Basic concepts of IEC 61499. In *Verteilte Automatisierung
          - Modelle und Methoden für Entwurf, Verifikation, Engineering und Instru-
          mentierung (VA2000), Otto-von-Guericke-Universität Magdeburg*, pages 55–
          62, March 2000.

[CL11]   Yu Feng Chen and Zhi Wu Li. Design of a maximally permissive liveness-
          enforcing supervisor with a compressed supervisory structure for flexible man-
          ufacturing systems. *Automatica*, 47(5):1028–1034, May 2011.

[CW10]   K. Cai and W.M. Wonham. Supervisor localization: A top-down approach
          to distributed control of discrete-event systems. *IEEE Transactions on Auto-
          matic Control*, 55(6):605 – 618, March 2010.

[DA10]   A. Dideban and H. Alla. Feedback control logic synthesis for non safe petri
          nets. *CoRR*, abs/1003.4905, 2010.

[DHJ+04] J. Desel, H.-M. Hanisch, G. Juhás, R. Lorenz, and C. Neumair. *A Guide
          to Modelling and Control with Modules of Signal Nets.*, volume 3147/2004.
          Springer Berlin / Heidelberg, 2004.

[DM07]   V. Dubinin and D. Missal. Reverse partially-marked save net condition/event
          systems and its interpretation. In *Proceedings of the 6th All-Russian scientific*

-*practical Conference with the international participation "Contemporary Information Technologies in Science, Education and Practice"*, pages 168–191, Orenburg, Russia, November 2007.

[DR98]   J. Desel and W. Reisig. *Lectures on Petri Nets I: Basic Models, LNCS 1491*, chapter Place/transition Petri nets, pages 122 – 173. Springer-Verlag, 1998.

[DV06]   V. Dubinin and V. Vyatkin. Towards a formal semantics of IEC1499 function blocks. In *Proceedings of the 4th IEEE Conference on Industrial Informatics (INDIN2006)*, Singapore, 2006.

[Fis07]   Paul A. Fishwick, editor. *Handbook of Dynamic System Modeling*. Chapman & Hall/CRC Computer & Information Science Series. Chapman & Hall/CRC, 2007.

[Fok07]   Wan Fokkink. *Modelling Distributed Systems*. Texts in Theoretical Computer Science. An EATCS Series. Springer Berlin-Heidelberg, 2007.

[Gua00]   X. Guan. *Distributed Supervisory Control of Forbidden Conditions, and Automated Synthesis and Composition of Task Controllers*. PhD thesis, The Graduate School University of Kentucky, 2000.

[GV03]   Claude Girault and Rudiger Valk, editors. *Petri Nets for System Engineering: A Guide to Modeling, Verification, and Applications*. Springer-Verlag New York, Inc., Secaucus, NJ, USA, 2003.

[GX05]   A. Giua and X. Xie. Nonblocking control of Petri nets using unfolding. In *Proceedings of the 16th IFAC World Congress*, 2005.

[Han97]   A. Lüder; U. Anderssohn; H.-M. Hanisch. Logic controller synthesis for net condition / event systems with forbidden paths. In *European Control Conference 1997*, Brussels, Belgium, July 1997.

[HCdQ+10] R. C. Hill, J. E. R. Cury, M. H. de Queiroz, D. M. Tilbury, and S. Lafortune. Multi-level hierarchical interface-based supervisory control. *Automatica*, 46(7):1152–1164, 2010.

[HK90]   L. E. Holloway and B. H. Krogh. Synthesis of feedback control logic for a class of controlled Petri nets. *IEEE Trans. Autom. Control*, 35(5):514–523, May 1990. NewsletterInfo: 38.

[HL98]   H.-M. Hanisch and A. Lüder. Modular plant modelling technique and related controller synthesis problems. In *IEEE International Conference on Systems, Man, and Cybernetics*, pages 686–691, October 1998.

[HLR96]   H.-M. Hanisch, A. Lüder, and M. Rausch. Controller synthesis for net condition/event systems with incomplete state observation. In *Computer Integrated Manufacturing and Automation Technology (CIMAT 96)*, pages 351–356, Mai 1996.

[HR95]   H.-M. Hanisch and M. Rausch. Net condition/event systems with multiple condition outputs. In *ETFA Emerging Technologies and Factory Automation*, pages 592–600, Paris, France, October 1995.

[HTL97]   H.-M. Hanisch, J. Thieme, and A. Lüder. Towards a synthesis method for distributed safety controllers based on net condition/event systems. In *Journal of Intelligent Manufacturing*, pages 357–368. Chapman & Hall, (8) 1997.

[HVL94]   A. Haji-Valizadeh and K. A. Loparo. Decentralized supervisory predicate control of discrete event dynamical systems. In *Proceedings of the American Control Conference*, pages 1099–1103, June 1994.

[IA03a]   M. V. Iordache and P. J. Antsaklis. Admissible decentralized control of Petri nets. In *Proceedings of the American Control Conference, Denver, Colorado*. IEEE, June 2003.

[IA03b]   M. V. Iordache and P. J. Antsaklis. Decentralized control of Petri nets with constraint transformations. In *Proceedings of the American Control Conference, Denver, Colorado*. IEEE, June 2003.

[IA06]   M.V. Iordache and P.J. Antsaklis. *Supervisory Control of Concurrent Systems*. Systems and Control: Foundations & Applications. Birkhäuser Boston, 2006.

[JCK01]   S. Jaing, V. Chandra, and R. Kumar. Decentralized control of discrete event systems with multiple local specializations. In *Proceedings of 2001 American Control Conference, Part B*, June 2001.

[Jen97]   Kurth Jensen. *Basic Concepts, Analysis Methods and Practical Use*, volume 3 of *Monographs in Theoretical Computer Science. An EATCS Series*. Springer, Berlin, 1997.

[JN04]   R. Juhás, G. amd Lorenz and C. Neumair. Modelling and control with modules of signal nets. In *Lectures on Concurrency and Petri Nets from the 4th Advanced Course on Petri Nets (ACPN), LNCS 3098*, pages 585–625, 2004.

[Kar07]   Andrei Karatkevich. *Dynamic Analysis of Petri Net-Based Discrete Systems*, volume 356 of *Lecture Notes in Control and Information Sciences*. Springer Berlin, 2007.

[Kar09]   Sirko Karras. *Systematischer modellgestützter Entwurf von Steuerungen für Fertigungssysteme.* PhD thesis, Martin Luther University Halle-Wittenberg, Faculty of Engeneering Sciences, 2009.

[KHG97]   B. H. Krogh, L.E. Holloway, and A. Giua. A survey of Petri net methods for controlled discrete event systems. In *Journal of Discrete Event Dynamical Systems, 7(2),* 1997.

[KK96]   B. H. Krogh and S. Kowalewski. Sate feedback control of condition/event systems. *Mathematical and computer modelling (Math. comput. model.),* 23(11-12):161–173, 1996.

[KMP06]   J. Komenda, H. March, and S. Pinchinat. A constructive and modular approach to decentralized supervisory control problems. In *3rd IFAC Workshop on Discrete-Event System Design, DESD 06, Rydzyna,* 2006.

[Kro87]   B. H. Krogh. Controlled petri nets and maximally permissive feedback logik. In *Proceedings of 25th Annual Allerton Conference, Urbana, IL, USA,* September 1987.

[KvSGM05]   J. Komenda, J. van Schuppen, B. Gaudin, and H. Marchand. Modular supervisory control with general indecomposable specification languages. In *Proceedings of the Conference on Decision on Control,* 2005.

[LBLW01]   R.L. Leduc, B.A. Brandin, M. Lawford, and W.M. Wonham. Hierarchical interface-based supervisory control: Serial case. In *Proceedings of the 40th Conference in Decision and Control,* Orlando, USA., 2001.

[LD07]   R.J. Leduc and P. Dai. Synthesis method for hierarchical interface-based supervisory control. In *Proc. of 26th American Control Conference,* pages 4260–4267, New York City, USA, July 2007.

[LHR96]   A. Lüder, H.-M. Hanisch, and M. Rausch. Synthesis of forcing/locking controllers based on net condition/event systems. In *IEEE International Conference on Emerging Technologies and Factory Automation(EFTA '96), Volume 2,* pages 341–347, Kauai, Hawaii, November 1996. IEEE.

[Lüd00]   A. Lüder. *Formaler Steuerungsentwurf mit modularen diskreten Verhaltensmodellen.* PhD thesis, Martin-Luther-Universität Halle-Wittenberg, 2000.

[LW88]   F. Lin and W. M. Wonham. Decentralized supervisory control of discrete-event systems. *Information Sciences,* 44:199–224, 1988.

[LZ09]   Zhi Wu Li and Meng Chu Zhou. *Deadlock Resolution in Automated Manufacturing Systems: A Novel Petri Net Approach*. Springer Publishing Company, Incorporated, 1st edition, 2009.

[MA00]   J. O. Moody and P. J. Antsaklis. Petri net supervisory for des with uncontrollable and unobservable transitions. In *IEEE Trans. Automat. Contr., 45(3)*, pages 462–476. IEEE, 2000.

[Mar03]   Alan B. Marcovitz. *Introduction to Logic Design*. Mcgraw-Hill College, 2nd edition, 2003.

[MH06b]   D. Missal and H.-M. Hanisch. Synthesis of distributed controllers by means of a monolithic approach. In *Proceedings of the 11th IEEE Int. Conf. on Emerging Technologies and Factory Automation (ETFA'2006)*, pages 356–363, Prague, Czech Republic, September 2006.

[MH07]   D. Missal and H.-M. Hanisch. Modular plant modelling for distributed control. In *IEEE Conference on Systems, Man, and Cybernetics*, pages 3475–3480, Montreal, Canada, October 2007.

[MH08a]   D. Missal and H.-M. Hanisch. Maximally permissive distributed safety control synthesis. 2008.

[MH08c]   D. Missal and H.-M. Hanisch. A modular synthesis approach for distributed safety controllers, part b: Modular control synthesis. In *17th IFAC World Congress, proceedings*, pages 14479–14484, Seoul, Korea, July 2008.

[MH09]   D. Missal and H.-M. Hanisch. Synthesis of distributed safety controllers with incomplete state observation. In *Annual Conference of the IEEE Industrial Electronics Society (IECONt2009)*, Porto, Portugal, 2009. (accepted).

[MS82]   J. Martinez and M. Silva. A simple and fast algorithm to obtain all invariants of a generalized Petri net. In *Selected Papers from the First and the Second European Workshop on Application and Theory of Petri Nets*, pages 301–310, London, UK, 1982. Springer-Verlag.

[PGH10]   S. Preusse, C. Gerber, and H.-M. Hanisch. Virtual start-up of plants using formal methods. *International Journal of Modelling, Identification and Control (IJMIC)*, submitted for publication, March 2010.

[PMG+10]   S. Preusse, D. Missal, C. Gerber, M. Hirsch, and H.-M. Hanisch. On the use of model-based IEC 61499 controller design. *International Journal of Discrete Event Control Systems (IJDECS)*, 1, 2010.

[RL03b] K. Rohloff and S. Lafortune. On the synthesis of safe control policies in decentralized control of discrete event systems. *IEEE Transactions on Automatic Control*, 48(6):1064–1068, June 2003.

[RR98] Wolfgang Reisig and Grzegorz Rozenberg, editors. *Lectures on Petri Nets II: Applications*, volume 1492 of *Lecture Notes in Computer Science*. Springer-Verlag Berlin-Heidelberg, 1998.

[RvS05] K.R Rohloff and J.H. van Schuppen. Approximation minimal communicated event sets for decentralized supervisory control. In *Proceedings of the IFAC World Congress*, Prague, Czech Republic, July 2005.

[RW86] P. J. Ramadge and W. M. Wonham. Modular supervisory control for discrete event systems. In *Seventh Internat. Conf. Analysis and Optimization of Systems, Nice, France*, June 1986.

[RW87a] P. J. Ramadge and W. M. Wonham. Modular feedback logic for discrete event systems. *SIAM J. Control Optim.*, 25(5):1202–1218, 1987.

[RW87b] P.J. Ramadge and W. M. Wonham. Supervisory control of a class of discrete event processes. *SIAM Journal of Control and Optimization*, 25:206–230, 1987.

[RW89] P. J. G. Ramadge and W. M. Wonham. The control of discrete event systems. In *Proceedings of the IEEE*, pages 81–97. IEEE, IEEE, January 1989.

[RW05] J.H. Richter and F. Wenck. Hierarchical interface-based supervisory control of a bottling plant. In *Proceedings of the 16th IFAC World Congress*, Prague, Czech Republic, 2005.

[SB11] K. Schmidt and C. Breindl. Maximally permissive hierarchical control of decentralized discrete event systems. *IEEE Transactions on Automatic Control*, 56(4):723–737, April 2011.

[SC01] Hessam S. Sarjoughian and Francois E. Cellier, editors. *Discrete Event Modeling and Simulation Technologies: A Tapestry of Systems and AI-Based Theories and Methodologies*. Springer-Verlag New York, Inc., New York, NY, USA, 2001.

[Shi98] Sajjan G. Shiva. *Introduction to Logic Design*. Marcel Dekker Inc., 2nd edition, 1998.

[SR02] P.H. Starke and St. Roch. Analysing signal-net systems. Technical report, Humboldt-Universität zu Berlin, Berlin, September 2002.

[Sre96]   R. S. Sreenivas. On asymptotical efficient solutions for a class of supervisory
           control problems. *IEEE Transactions on Automatic Control*, 41(12):1736–
           1750, December 1996.

[Sta90]   P. H. Starke, editor. *Analyse von Petri-Netz-Modellen*. B. G. Teubner
           Stuttgart, 1990.

[SZC+06a] Ch. Suender, A. Zoitl, J. H. Christensen, V. Vyatkin, R. W. Brennan, A. Valen-
           tini, L. Ferrarini, Th. Strasser, J. L. Martinez-Lastra, and F. Auinger. Usabil-
           ity and interoperability of IEC 61499 based distributed automation systems.
           In *International Conference on Industrial Informatics (INDIN)*, pages 31–37,
           Singapore, 2006.

[SZC+06b] C. Sunder, A. Zoitl, J.-H. Christensen, V. Vyatkin, R.-W. Brennan, A. Valen-
           tini, L. Ferrarini T. Strasser, J.-L Martinez-Lastra, and F. Auinger. Usabil-
           ity and interoperability of IEC 61499 based distributed automations systems.
           In *IEEE International Conference on Industrial Informatics (INDIN 2006)*,
           pages 31–37, August 2006.

[Thi02]   J. Thieme. *Symbolische Erreichbarkeitsanalyse und automatische Implemen-
           tierung strukturierter, zeitbewerterter Steuerungsmodelle*. Hallenser Schriften
           zur Automatisierungtechnik. Logos-Verl., 2002.

[Tri04b]   S. Tripakis. Undecidable problems of decentralized observation and control
           for regular languages. In *Information Processing Letters, Volume 90 Issue 1*,
           pages 21–28. April 2004.

[UZ06]   M. Uzam and M. C. Zhou. An improved iterative syhthesis method for liveness
           enforcing supervisors of flexible manufacturing systems. *International Journal
           of Production Research*, 44(10):1987Ü2030, 2006.

[Val98]   A. Valmari. *Lectures on Petri Nets I: Basic Models, LNCS 1491*, chapter The
           State Explosion Problem, pages 429–528. Springer-Verlag, 1998.

[VH99]   V. Vyatkin and H.-M. Hanisch. A modeling approach for verification of
           IEC1499 function blocks using net condition / event systems. In *Emerging
           Technologies and Factory Automation (ETFA'99), proceedings*, pages 261–269,
           Catalonia, Spain, October 1999.

[VH03]   V. Venugopal and L. E. Holloway. Coordinating concurrency to avoid for-
           bidden states in condition models. In *Emerging Technologies and Factory
           Automation, 2003. ETFA '03*, pages 90–97. IEEE, September 2003.

[VHK⁺06]   V. Vyatkin, H.-M. Hanisch, S. Karras, T. Pfeiffer, and V. Dubinin. Rapid engineering and re-configuration of automation objects using formal verification. *International Journal of Manufacturing Research*, 1(4):382–404, 2006.

[vS98]   J. H. van Schuppen. Decentralized supervisory control with information structures. In *Proceedings International Workshop on Discrete Event Systems (WODES98)*, pages 278–283. IEE Press, London, 1998.

[Vya06]   V. Vyatkin. *IEC 61499 Function Blocks for Embedded and Distributed Control Systems*. O3NEIDA - Instrumentation Society of America, 2006.

[WR84]   W. M. Wonham and P. J. Ramadge. On the supremal controllable sublanguage of a given language. In *Proceeding of the 23rd IEEE Conf. on Decision and Control, Las Vegas, Nevada*. IEEE, December 1984.

[YL02]   T.-S. Yoo and S. Lafortune. A general architecture for decentralized supervisory control of discrete-event systems. In *Journal of Discrete Event Dynamical Systems: Theory and Applications*, pages 335–377. 2002.

[YL04]   T.-S. Yoo and S Lafortune. Decentralized supervisory control with conditional decisions: supervisor existence. *IEEE Transactions on Automatic Control*, 49(11):1886–1904, Nov. 2004.

[YMAL96]   E. Yamalidou, J. O. Moody, P. J. Antsaklis, and M. D. Lemmon. Feedback control of Petri nets based on place invariants. In *Automatica, 32(1)*, pages 15–28. 1996.

[ZL10]   C.-F. Zhong and Z.-W. Li. Design of liveness-enforcing supervisors via transforming plant petri net models of FMS. *Asian Journal of Control*, 12:240Ũ252, 2010.

Index

Short Vita

Dirk Missal
born in 1977 in Halle (Saale), Germany

Employment History

March 2010 -	Dow Olefinverbund GmbH
	Advance Control Engineer at Engineering Solution - Process Automation
Nov. 2005 - Feb. 2010	Martin Luther University Halle-Wittenberg - Chair for Automation Technology
	scientific staff (third-party funded)

Education

Sep. 2000 - Nov. 2005	Martin Luther University Halle-Wittenberg, Department for Engineering Sciences
	degree program: Engineering Sciences/ Ingenieur- Informatik
	Degree: Diplom-Ingenieur
July 1996 - June 1999	Dow BSL Olefinverbund GmbH
	apprenticeship process control technician; degree: Facharbeiter
23.06.1996	A-level at the Suedstadt Gymnasium Halle(Saale)

Civilian Service

Aug. 1999 - June 2000	Max-Bürger Zentrum Berlin

Abstract

Modern control systems in manufacturing are characterized by rising complexity in size and functionality. They are highly decentralized and constitute a network of physically and functionally distributed controllers collaborating to perform the control tasks. That goes along with a further growing demand on safety and reliability. A distributed control architecture supporting functional decomposition of large systems as well as accommodating flexibility of modular systems is defined.

This work describes the formal synthesis of distributed control functions for the sub area of safety requirements. The formal synthesis is applied to avoid the potentially faulty influence of human work through the whole process from the formal specification to the executable control function. Starting points are a formal model of the uncontrolled plant behavior and a formal specification of forbidden behavior. The formulation of the specification and the modeling is exemplified on a manufacturing system in lab-scale.

The introduced synthesis methods produce controller models describing the correct control actions to achieve the given specification. The methods use symbolic backward search from a forbidden state to determine the last admissible state before entering an uncontrollable trajectory to a forbidden state. Hence, the determination of the reachable state space is avoided to reduce the computational complexity. The use of partial markings leads to a further reduction. The complexity is an important obstacle for the use of formal methods on real-scale applications. The monolithic synthesis approach is proven to result in maximally permissive results. The modular approach is not maximally permissive but the more efficient way to distributed control functions.

The implementation of the generated controller model as executable Function Blocks according to IEC61499 is addressed in the last part of this work. The distributed control predicates are embedded as structured text instruction into different interacting Function Block types according to the distributed control structure. This last step finalizes the sequence from a formal model and the specification to fully automatically-generated executable control code.